Survey of Text Mining

Michael W. Berry

Editor

Survey of Text Mining
Clustering, Classification, and Retrieval

 Springer

Michael W. Berry
Department of Computer Science
University of Tennessee
203 Claxton Complex
Knoxville, TN 37996-3450, USA
berry@cs.utk.edu

Cover illustration: Visualization of three major clusters in the *L.A. Times* news database when document vectors are projected into the 3-D subspace spanned by the three most relevant axes determined using COV rescale. This figure appears on p. 118 of the text.

Library of Congress Cataloging-in-Publication Data
Survey of text mining : clustering, classification, and retrieval / editor, Michael W. Berry.
    p. cm.
  Includes bibliographical references and index.
  ISBN 0-387-95563-1 (alk. Paper)
    1. Data mining—Congresses. 2. Cluster analysis—Congresses. 3. Discriminant analysis—Congresses. I. Berry, Michael W.

QA76.9.D343S69 2003
006.3—dc21   2003042434

ISBN 0-387-95563-1   Printed on acid-free paper.

Printed in the United States of America.   (EB)

9 8 7 6 5 4

springer.com

Contents

Preface   xi

Contributors   xiii

I   Clustering and Classification   1

1   Cluster-Preserving Dimension Reduction Methods for Efficient Classification of Text Data   3
Peg Howland and Haesun Park
    1.1   Introduction   3
    1.2   Dimension Reduction in the Vector Space Model   4
    1.3   A Method Based on an Orthogonal Basis of Centroids   5
        1.3.1   Relationship to a Method from Factor Analysis   7
    1.4   Discriminant Analysis and Its Extension for Text Data   8
        1.4.1   Generalized Singular Value Decomposition   9
        1.4.2   Extension of Discriminant Analysis   11
        1.4.3   Equivalence for Various S_1 and S_2   14
    1.5   Trace Optimization Using an Orthogonal Basis of Centroids . .   16
    1.6   Document Classification Experiments   17
    1.7   Conclusion   20
    References   22

2   Automatic Discovery of Similar Words   25
Pierre P. Senellart and Vincent D. Blondel
    2.1   Introduction   25
    2.2   Discovery of Similar Words from a Large Corpus   26
        2.2.1   A Document Vector Space Model   27
        2.2.2   A Thesaurus of Infrequent Words   28
        2.2.3   The SEXTANT System   29
        2.2.4   How to Deal with the Web   32
    2.3   Discovery of Similar Words in a Dictionary   33

	2.3.1	Introduction	33
	2.3.2	A Generalization of Kleinberg's Method	33
	2.3.3	Other Methods	35
	2.3.4	Dictionary Graph	36
	2.3.5	Results	37
	2.3.6	Future Perspectives	41
2.4	Conclusion	41
	References	42

3 Simultaneous Clustering and Dynamic Keyword Weighting for Text Documents   **45**

Hichem Frigui and Olfa Nasraoui

3.1	Introduction	45
3.2	Simultaneous Clustering and Term Weighting of Text Documents	47
3.3	Simultaneous Soft Clustering and Term Weighting of Text Documents	52
3.4	Robustness in the Presence of Noise Documents	56
3.5	Experimental Results	57
	3.5.1 Simulation Results on Four-Class Web Text Data	57
	3.5.2 Simulation Results on 20 Newsgroups Data	59
3.6	Conclusion	69
	References	70

4 Feature Selection and Document Clustering   **73**

Inderjit Dhillon, Jacob Kogan, and Charles Nicholas

4.1	Introduction	73
4.2	Clustering Algorithms	74
	4.2.1 Means Clustering Algorithm	74
	4.2.2 Principal Direction Divisive Partitioning	78
4.3	Data and Term Quality	80
4.4	Term Variance Quality	81
4.5	Same Context Terms	86
	4.5.1 Term Profiles	87
	4.5.2 Term Profile Quality	87
4.6	Spherical Principal Directions Divisive Partitioning	90
	4.6.1 Two-Cluster Partition of Vectors on the Unit Circle	90
	4.6.2 Clustering with sPDDP	96
4.7	Future Research	98
	References	99

II   Information Extraction and Retrieval   **101**

5   Vector Space Models for Search and Cluster Mining   **103**

Mei Kobayashi and Masaki Aono
5.1   Introduction   103
5.2   Vector Space Modeling (VSM)   105
    5.2.1   The Basic VSM Model for IR   105
    5.2.2   Latent Semantic Indexing (LSI)   107
    5.2.3   Covariance Matrix Analysis (COV)   108
    5.2.4   Comparison of LSI and COV   109
5.3   VSM for Major and Minor Cluster Discovery   111
    5.3.1   Clustering   111
    5.3.2   Iterative Rescaling: Ando's Algorithm   111
    5.3.3   Dynamic Rescaling of LSI   113
    5.3.4   Dynamic Rescaling of COV   114
5.4   Implementation Studies   115
    5.4.1   Implementations with Artificially Generated Datasets .   115
    5.4.2   Implementations with L.A. Times News Articles   118
5.5   Conclusions and Future Work   120
References   120

6   HotMiner: Discovering Hot Topics from Dirty Text   123
Malú Castellanos
6.1   Introduction   124
6.2   Related Work   128
6.3   Technical Description   130
    6.3.1   Preprocessing   130
    6.3.2   Clustering   132
    6.3.3   Postfiltering   133
    6.3.4   Labeling   136
6.4   Experimental Results   137
6.5   Technical Description   143
    6.5.1   Thesaurus Assistant   145
    6.5.2   Sentence Identifier   147
    6.5.3   Sentence Extractor   149
6.6   Experimental Results   151
6.7   Mining Case Excerpts for Hot Topics   153
6.8   Conclusions   154
References   155

**7   Combining Families of Information Retrieval Algorithms Using
Metalearning   159**
Michael Cornelson, Ed Greengrass, Robert L. Grossman, Ron Karidi,
and Daniel Shnidman
7.1   Introduction   159
7.2   Related Work   161
7.3   Information Retrieval   162
7.4   Metalearning   164

7.5   Implementation   166
7.6   Experimental Results   166
7.7   Further Work   167
7.8   Summary and Conclusion   168
References   168

III   Trend Detection   171

8   Trend and Behavior Detection from Web Queries   173
Peiling Wang, Jennifer Bownas, and Michael W. Berry

8.1   Introduction   173
8.2   Query Data and Analysis   174
       8.2.1   Descriptive Statistics of Web Queries   175
       8.2.2   Trend Analysis of Web Searching   176
8.3   Zipf's Law   178
       8.3.1   Natural Logarithm Transformations   178
       8.3.2   Piecewise Trendlines   179
8.4   Vocabulary Growth   179
8.5   Conclusions and Further Studies   181
References   182

9   A Survey of Emerging Trend Detection in Textual Data Mining   185
April Kontostathis, Leon M. Galitsky, William M. Pottenger, Soma
Roy, and Daniel J. Phelps

9.1   Introduction   186
9.2   ETD Systems   187
       9.2.1   Technology Opportunities Analysis (TOA)   189
       9.2.2   CIMEL: Constructive, Collaborative Inquiry-Based
                Multimedia E-Learning   191
       9.2.3   TimeMines   195
       9.2.4   New Event Detection   199
       9.2.5   ThemeRiver™   201
       9.2.6   PatentMiner   204
       9.2.7   HDDI™   207
       9.2.8   Other Related Work   211
9.3   Commercial Software Overview   211
       9.3.1   Autonomy   212
       9.3.2   SPSS LexiQuest   212
       9.3.3   ClearForest   213
9.4   Conclusions and Future Work   213
9.5   Industrial Counterpoint: Is ETD Useful?   Dr. Daniel J. Phelps,
       Leader, Information Mining Group, Eastman Kodak   215
References   219

Bibliography   **225**

Index   **241**

Preface

As we enter the second decade of the World Wide Web (WWW), the textual revolution has seen a tremendous change in the availability of online information. Finding information for just about any need has never been more automatic – just a keystroke or mouseclick away. While the digitalization and creation of textual materials continues at light speed, the ability to navigate, mine, or casually browse through documents too numerous to read (or print) lags far behind.

What approaches to text mining are available to efficiently organize, classify, label, and extract relevant information for today's information-centric users? What algorithms and software should be used to detect emerging trends from both text streams and archives? These are just a few of the important questions addressed at the Text Mining Workshop held on April 13, 2002 in Arlington, VA. This workshop, the second in a series of annual workshops on text mining, was held on the third day of the Second SIAM International Conference on Data Mining (April 11–13, 2002).

With close to 60 applied mathematicians and computer scientists representing universities, industrial corporations, and government laboratories, the workshop featured both invited and contributed talks on important topics such as efficient methods for document clustering, synonym extraction, efficient vector space models and metalearning approaches for text retrieval, hot topic discovery from dirty text, and trend detection from both queries and documents. The workshop was sponsored by the Army High Performance Computing Research Center (AHPCRC) – Laboratory for Advanced Computing, SPSS, Insightful Corporation, and Salford Systems.

Several of the invited and contributed papers presented at the 2002 Text Mining Workshop have been compiled and expanded for this volume. Collectively, they span several major topic areas in text mining:

   I. Clustering and Classification,

   II. Information Extraction and Retrieval, and

   III. Trend Detection.

In Part I (Clustering and Classification), Howland and Park present cluster-preserving dimension reduction methods for efficient text classification; Senellart and Blondel demonstrate thesaurus construction using similarity measures between vertices in graphs; Frigui and Nasraoui discuss clustering and keyword weighting; and Dhillon, Kogan, and Nicholas illustrate how both feature selection and document clustering can be accomplished with reduced dimension vector space models.

In Part II (Information Extraction and Retrieval), Kobayashi and Aono demonstrate the importance of detecting and interpreting minor document clusters using a vector space model based on Principal Component Analysis (PCA) rather than the popular Latent Semantic Indexing (LSI) method; Castellanos demonstrates how important topics can be extracted from dirty text associated with search logs in the customer support domain; and Cornelson et al. describe an innovative approach to information retrieval based on metalearning in which several algorithms are applied to the same corpus.

In Part III (Trend Detection), Wang, Bownas, and Berry mine Web queries from a university website in order to expose the type and nature of query characteristics through time; and Kontostathis et al. formally evaluate available Emerging Trend Detection (ETD) systems and discuss future criteria for the development of effective industrial-strength ETD systems.

Each chapter of this volume is preceded by a brief chapter overview and concluded by a list of references cited in that chapter. A main bibliography of all references cited and a subject-level index are also provided at the end of the volume. This volume details state-of-the-art algorithms and software for text mining from both the academic and industrial perspectives. Familiarity or coursework (undergraduate-level) in vector calculus and linear algebra is needed for several of the chapters in Parts I and II. While many open research questions still remain, this collection serves as an important benchmark in the development of both current and future approaches to mining textual information.

Acknowledgments: The editor would like to thank Justin Giles, Kevin Heinrich, and Svetlana Mironova who were extremely helpful in proofreading many of the chapters of this volume. Justin Giles also did a phenomenal job in managing all the correspondences with the contributing authors.

Michael W. Berry
Knoxville, TN
December 2002

Contributors

Masaki Aono
IBM Research, Tokyo Research Laboratory
1623-14 Shimotsuruma, Yamato-shi
Kanagawa-ken 242-8502
Japan
Email:aono@jp.ibm.com

Michael W. Berry
Department of Computer Science
University of Tennessee
203 Claxton Complex
Knoxville, TN 37996-3450
Email: berry@cs.utk.edu
Homepage: http://www.cs.utk.edu/~berry

Vincent D. Blondel
Division of Applied Mathematics
Université de Louvain
4, Avenue Georges Lemaître
B-1348 Louvain-la-neuve
Belgium
Email: blondel@inma.ucl.ac.be
Homepage: http://www.inma.ucl.ac.be/~blondel

Jennifer Bownas
School of Information Sciences
University of Tennessee
451 Communications Building
1345 Circle Park Drive
Knoxville, TN 37996-0341
Email: jbownas@.utk.edu

Malú Castellanos
IETL Department
Hewlett-Packard Laboratories
1501 Page Mill Road MS-1148
Palo Alto, CA 94304
Email: castella@hpl.hp.com

Michael Cornelson
Arista Technology Group
1332 West Fillmore Street
Chicago, IL 60607
Email: mcornel@aristainc.com

Inderjit S. Dhillon
Department of Computer Sciences
University of Texas
ACES 2.332
201 East 24th Street
Austin, TX 78712-1188
Email: inderjit@cs.utexas.edu
Homepage: http://www.cs.utexas.edu/users/inderjit

Hichem Frigui
Department of Electrical and Computer Engineering
206 Engineering Science Building
University of Memphis
Memphis, TN 38152-3180
Email: hfrigui@memphis.edu
Homepage: http://prlab.ee.memphis.edu/frigui

Leon M. Galitsky
Department of Computer Science and Engineering
Lehigh University
19 Memorial Drive West
Bethlehem, PA 18015

Ed Greengrass
United States Department of Defense
9800 Savage Road
Fort Meade, MD 20755-6000
Attn: R521
Email: edgreen@afterlife.ncsc.mil

Robert Grossman
   Department of Mathematics, Statistics, and Computer Science
   Mail Code 249
   University of Illinois at Chicago
   851 S. Morgan Street
   Chicago, IL 60607
   Email: grossman@uic.edu
   Homepage: http://www.lac.uic.edu/~grossman/index.htm

Peg Howland
   Department of Computer Science and Engineering
   University of Minnesota
   4-192 EE/CS Building
   Minneapolis, MN 55455
   Email: howland@cs.umn.edu

Ron Karidi
   LivePerson Inc.
   462 Seventh Avenue, 21st Floor
   New York, NY 10018
   Email: ron@liveperson.com

Mei Kobayashi
   IBM Research, Tokyo Research Laboratory
   1623-14 Shimotsuruma, Yamato-shi
   Kanagawa-ken 242-8502
   Japan
   Email:mei@jp.ibm.com

Jacob Kogan
   Department of Mathematics and Statistics
   University of Maryland, Baltimore County
   1000 Hilltop Circle
   Baltimore, MD 21250
   Email: kogan@math.umbc.edu
   Homepage: http://www.math.umbc.edu/~kogan

April Kontostathis
   Department of Computer Science and Engineering
   Lehigh University
   19 Memorial Drive West
   Bethlehem, PA 18015
   Email: apk5@lehigh.edu

Olfa Nasraoui
   Department of Electrical and Computer Engineering
   206 Engineering Science Building
   University of Memphis
   Memphis, TN 38152-3180
   Email: onasraou@memphis.edu
   Homepage: http://www.ee.memphis.edu/people/faculty/nasraoui/nasraoui.html

Charles Nicholas
   Department of Computer Science and Electrical Engineering
   University of Maryland, Baltimore County
   1000 Hilltop Circle
   Baltimore, MD 21250
   Email: nicholas@cs.umbc.edu
   Homepage: http://www.cs.umbc.edu/~nicholas

Haesun Park
   Department of Computer Science and Engineering
   University of Minnesota
   4-192 EE/CS Building
   Minneapolis, MN 55455
   Email: hpark@cs.umn.edu
   Homepage: http://www-users.cs.umn.edu/~hpark

Daniel J. Phelps
   Eastman Kodak Research and Development Laboratories
   1999 Lake Avenue
   Rochester, NY 14650-2138

William M. Pottenger
   Department of Computer Science and Engineering
   Lehigh University
   19 Memorial Drive West
   Bethlehem, PA 18015
   Email: pottenger@cse.lehigh.edu
   Homepage: http://www.cse.lehigh.edu/~billp

Soma Roy
   Department of Computer Science and Engineering
   Lehigh University
   19 Memorial Drive West
   Bethlehem, PA 18015
   Email: sor4@lehigh.edu

Pierre Senellart
    École Normale Supérieure
    45, rue d'Ulm
    F-75005 Paris
    France
    Email: pierre.senellart@ens.fr
    Homepage: http://www.eleves.ens.fr/home/senellar

Daniel Shnidman
    Baesch Computer Consulting, Inc.
    510 McCormick Drive
    Suite A
    Glen Burnie, MD 21061
    Email: dans@bcci.com

Peiling Wang
    School of Information Sciences
    University of Tennessee
    451 Communications Building
    1345 Circle Park Drive
    Knoxville, TN 37996-0341
    Email: peilingw@.utk.edu
    Homepage: http://www.sis.utk.edu/faculty/wang

Part I

Clustering and Classification

1

Cluster-Preserving Dimension Reduction Methods for Efficient Classification of Text Data

Peg Howland
Haesun Park

Overview

In today's vector space information retrieval systems, dimension reduction is imperative for efficiently manipulating the massive quantity of data. To be useful, this lower-dimensional representation must be a good approximation of the original document set given in its full space. Toward that end, we present mathematical models, based on optimization and a general matrix rank reduction formula, which incorporate a priori knowledge of the existing structure. From these models, we develop new methods for dimension reduction based on the centroids of data clusters. We also adapt and extend the discriminant analysis projection, which is well known in pattern recognition. The result is a generalization of discriminant analysis that can be applied regardless of the relative dimensions of the term-document matrix.

We illustrate the effectiveness of each method with document classification results from the reduced representation. After establishing relationships among the solutions obtained by the various methods, we conclude with a discussion of their relative accuracy and complexity.

1.1   Introduction

The vector space information retrieval system, originated by Gerard Salton [Sal71, SM83], represents documents as vectors in a vector space. The document set comprises an $m \times n$ term-document matrix A, in which each column represents a document, and each entry $A(i, j)$ represents the weighted frequency of term i in document j. A major benefit of this representation is that the algebraic structure of the vector space can be exploited [BDO95]. To achieve higher efficiency in manipulating the data, it is often necessary to reduce the dimension dramatically. Especially when the data set is huge, we can assume that the data have a cluster structure, and it is often necessary to cluster the data [DHS01] first to utilize the tremendous amount of information in an efficient way. Once the columns of A are grouped into clusters, rather than treating each column equally regardless of its membership in a specific cluster, as is done in the singular value decomposition (SVD) [GV96], the dimension reduction methods we discuss attempt to preserve this information.

These methods also differ from probability and frequency-based methods, in which a set of representative words is chosen. For each dimension in the reduced space we cannot easily attach corresponding words or a meaning. Each method attempts to choose a projection to the reduced dimension that will capture a priori knowledge of the data collection as much as possible. This is important in information retrieval, since the lower rank approximation is not just a tool for rephrasing a given problem into another one which is easier to solve [HMH00], but the reduced representation itself will be used extensively in further processing of data.

With that in mind, we observe that dimension reduction is only a preprocessing stage. Even if this stage is a little expensive, it may be worthwhile if it effectively reduces the cost of the postprocessing involved in classification and document retrieval, which will be the dominating parts computationally. Our experimental results illustrate the trade-off in effectiveness versus efficiency of the methods, so that their potential application can be evaluated.

1.2   Dimension Reduction in the Vector Space Model

Given a term-document matrix

$$A = [a_1 \quad a_2 \quad \cdots \quad a_n] \in \mathbb{R}^{m \times n},$$

the problem is to find a transformation that maps each document vector a_j in the m-dimensional space to a vector y_j in the l-dimensional space for some $l < m$:

$$a_j \in \mathbb{R}^{m \times 1} \rightarrow y_j \in \mathbb{R}^{l \times 1}, \quad 1 \le j \le n.$$

The approach we discuss in Section 1.4 computes the transformation directly from A. Rather than looking for the mapping that achieves this explicitly, another approach rephrases this as an approximation problem where the given matrix A is decomposed into two matrices B and Y as

$$A \approx BY, \tag{1.1}$$

where both $B \in \mathbb{R}^{m \times l}$ with rank$(B) = l$ and $Y \in \mathbb{R}^{l \times n}$ with rank$(Y) = l$ are to be found. This lower rank approximation is not unique since for any nonsingular matrix $Z \in \mathbb{R}^{l \times l}$,

$$A \approx BY = (BZ)(Z^{-1}Y),$$

where $rank(BZ) = l$ and $rank(Z^{-1}Y) = l$. This problem of approximate decomposition (1.1) can be recast in two different but related ways. The first is in terms of a matrix rank reduction formula and the second is as a minimization problem. A matrix rank reduction formula that has been studied extensively in both numerical linear algebra [CF79, CFG95] and applied statistics/psychometrics [Gut57, HMH00] is summarized here.

THEOREM 1.1 *(Matrix Rank Reduction Theorem) Let $A \in \mathbb{R}^{m \times n}$ be a given matrix with $rank(A) = r$. Then the matrix*

$$E = A - (AP_2)(P_1AP_2)^{-1}(P_1A), \tag{1.2}$$

where $P_1 \in \mathbb{R}^{l \times m}$ and $P_2 \in \mathbb{R}^{n \times l}$, $l \leq r$, satisfies

$$rank(E) = rank(A) - rank((AP_2)(P_1AP_2)^{-1}(P_1A)) \tag{1.3}$$

if and only if $P_1AP_2 \in \mathbb{R}^{l \times l}$ is nonsingular.

The only restrictions on the factors P_1 and P_2 are on their dimensions, and that the product P_1AP_2 be nonsingular. It is this choice of P_1 and P_2 that makes the dimension reduction flexible and incorporation of a priori knowledge possible. In fact, in [CFG95] it is shown that many fundamental matrix decompositions can be derived using this matrix rank reduction formula. Letting

$$BY = (AP_2)(P_1AP_2)^{-1}P_1A,$$

we see that minimizing the error matrix E in some p-norm is equivalent to solving the problem

$$\min_{B,Y} ||A - BY||_p. \tag{1.4}$$

The incorporation of a priori knowledge can be translated into choosing the factors P_1 and P_2 in (1.2) or adding a constraint in the minimization problem (1.4). However, mathematical formulation of this knowledge as a constraint is not always easy. In the next section, we discuss ways to choose the factors B and Y so that knowledge of the clusters from the full dimension is reflected in the dimension reduction.

1.3   A Method Based on an Orthogonal Basis of Centroids

For simplicity of discussion, we assume that the columns of A are grouped into k clusters as

$$A = [A_1 \quad A_2 \quad \cdots \quad A_k] \quad \text{where} \quad A_i \in \mathbb{R}^{m \times n_i}, \quad \text{and} \quad \sum_{i=1}^{k} n_i = n. \tag{1.5}$$

Let N_i denote the set of column indices that belong to cluster A_i. The centroid $c^{(i)}$ of each cluster is computed by taking the average of the columns in A_i; that is,

$$c^{(i)} = \frac{1}{n_i} \sum_{j \in N_i} a_j,$$

and the global centroid c is defined as

$$c = \frac{1}{n} \sum_{j=1}^{n} a_j.$$

The centroid vector achieves the minimum variance in the sense:

$$\sum_{j=1}^{n} \|a_j - c\|_2^2 = \min_{x \in \mathbb{R}^{n \times 1}} \sum_{j=1}^{n} \|a_j - x\|_2^2 = \min_{x \in \mathbb{R}^{n \times 1}} \|A - xe^T\|_F^2,$$

where $e = (1, \cdots, 1)^T \in \mathbb{R}^{n \times 1}$. Applying this within each cluster, we can find one vector to represent the entire cluster. This suggests that we choose the columns of B in the minimization problem (1.4) to be the centroids of the k clusters, and then solve the least squares problem [Bjö96]

$$\min_{Y \in \mathbb{R}^{k \times n}} \|A - CY\|_F,$$

where

$$C = [c^{(1)} \quad c^{(2)} \quad \cdots \quad c^{(k)}].$$

Note that for this method, the reduced dimension l is the same as the number of clusters k.

To express this in terms of the matrix rank reduction formula, we define a grouping matrix $H \in \mathbb{R}^{n \times k}$ as

$$H = F \cdot (F^T F)^{-1}, \tag{1.6}$$

where $\quad F \in \mathbb{R}^{n \times k} \quad$ and $\quad F(i, j) = \begin{cases} 1 & \text{if document } i \text{ belongs to cluster } j, \\ 0 & \text{otherwise.} \end{cases}$

It is easy to see that the matrix C can be written as $C = AH$. In addition, the solution is $Y = (C^T C)^{-1} C^T A$, which in turn yields the matrix rank reduction expression (1.2)

$$\begin{aligned} E &= A - CY \\ &= A - (AH)(C^T C)^{-1} C^T A \\ &= A - (AH)(H^T A^T AH)^{-1}(H^T A^T A). \end{aligned}$$

This shows that the prefactor is $P_1 = H^T A^T$ and the postfactor is $P_2 = H$ in the Centroid method of Park et al. [PJR03].

Algorithm 1.1 CentroidQR

Given a data matrix $A \in \mathbb{R}^{m \times n}$ with k clusters, compute a k-dimensional representation Y of A.

1. Compute the centroid $c^{(i)}$ of the ith cluster, $1 \leq i \leq k$.

2. Set $C = [c^{(1)} \quad c^{(2)} \quad \cdots \quad c^{(k)}]$.

3. Compute the reduced QR decomposition of C, which is $C = Q_k R$.

4. Solve $\min_Y \| Q_k Y - A \|_F$ (in fact, $Y = Q_k^T A$).

Algorithm 1.2 Multiple Group

Given a data matrix $A \in \mathbb{R}^{m \times n}$ with k clusters, compute a k-dimensional representation Y of A.

1. Compute the matrix $W = A^T A H$, where H is the grouping matrix defined in Eq. (1.6).

2. Compute $S = H^T W$.

3. Compute the Cholesky factor T of S, so that $T T^T = S$.

4. $Y = T^{-1} W^T$.

If the factor B in (1.4) has orthonormal columns, then the matrix Y itself gives a good approximation of A in terms of their correlations:

$$A^T A \approx Y^T B^T B Y = Y^T Y \quad \text{when } B^T B = I.$$

For a given matrix B, this can be achieved by computing its reduced QR decomposition [GV96]. When $B = C$, the result is the CentroidQR method, which is presented in Algorithm 1.1. For details of its development and properties, see [PJR03]. In Section 1.5, we show that the CentroidQR method solves a trace optimization problem, thus providing a link between the methods of discriminant analysis and those based on centroids.

1.3.1   Relationship to a Method from Factor Analysis

Before moving on to the subject of discriminant analysis, we establish the mathematical equivalence of the CentroidQR method to a method known in applied statistics/psychometrics for more than 50 years. In his book on factor analysis [Hor65], Horst attributes the multiple group method to Thurstone [Thu45]. We restate it in Algorithm 1.2, using the notation of numerical linear algebra.

In comparison, the solution from CentroidQR is given by

$$
\begin{aligned}
Y &= Q_k^T A \\
&= (A H R^{-1})^T A \\
&= R^{-T} H^T A^T A.
\end{aligned}
$$

From the uniqueness of the Cholesky factor, this matches the solution given in Algorithm 1.2, provided the CentroidQR method computes R as upper triangular with positive diagonal entries.

1.4   Discriminant Analysis and Its Extension for Text Data

The goal of discriminant analysis is to combine features of the original data in a way that most effectively discriminates between classes. With an appropriate extension, it can be applied to our goal of reducing the dimension of a term-document matrix in a way that most effectively preserves its cluster structure. That is, we want to find a linear transformation G^T that maps the m-dimensional document vector a_j into an l-dimensional vector as follows,

$$
G^T \in \mathbb{R}^{l \times m} : a_j \in \mathbb{R}^{m \times 1} \rightarrow y_j \in \mathbb{R}^{l \times 1}, \quad 1 \le j \le n.
$$

Assuming that the given data are already clustered, we seek a transformation that optimally preserves this cluster structure in the reduced dimensional space.

For this purpose, we first need to formulate a measure of cluster quality. When cluster quality is high, each cluster is tightly grouped, but well separated from the other clusters. To quantify this, scatter matrices are defined in discriminant analysis [Fuk90, TK99]. In terms of the centroids defined in the previous section, the within-cluster, between-cluster, and mixture scatter matrices are defined as

$$
S_w = \sum_{i=1}^{k} \sum_{j \in N_i} (a_j - c^{(i)})(a_j - c^{(i)})^T,
$$

$$
S_b = \sum_{i=1}^{k} \sum_{j \in N_i} (c^{(i)} - c)(c^{(i)} - c)^T = \sum_{i=1}^{k} n_i (c^{(i)} - c)(c^{(i)} - c)^T, \text{ and}
$$

$$
S_m = \sum_{j=1}^{n} (a_j - c)(a_j - c)^T,
$$

respectively. It is easy to show [JD88] that the scatter matrices have the relationship

$$
S_m = S_w + S_b. \tag{1.7}
$$

Applying G^T to A transforms the scatter matrices to

$$
S_w^Y = G^T S_w G, \ S_b^Y = G^T S_b G, \text{ and } S_m^Y = G^T S_m G,
$$

where the superscript Y denotes values in the l-dimensional space.

There are several measures of cluster quality that involve the three scatter matrices [Fuk90, TK99]. Since

$$\text{trace}(S_w) = \sum_{i=1}^{k} \sum_{j \in N_i} (a_j - c^{(i)})^T (a_j - c^{(i)}) = \sum_{i=1}^{k} \sum_{j \in N_i} \|a_j - c^{(i)}\|_2^2$$

measures the closeness of the columns within the clusters, and

$$\text{trace}(S_b) = \sum_{i=1}^{k} \sum_{j \in N_i} (c^{(i)} - c)^T (c^{(i)} - c) = \sum_{i=1}^{k} \sum_{j \in N_i} \|c^{(i)} - c\|_2^2$$

measures the separation between clusters, an optimal transformation that preserves the given cluster structure would maximize $\text{trace}(S_b^Y)$ and minimize $\text{trace}(S_w^Y)$.

This simultaneous optimization can be approximated by finding a transformation G that maximizes $\text{trace}((S_w^Y)^{-1} S_b^Y)$. However, this criterion cannot be applied when the matrix S_w is singular. In handling document data, it is often the case that the number of terms in the document collection is larger than the total number of documents (i.e., $m > n$ in the term-document matrix A), and therefore the matrix S_w is singular. Furthermore, in applications where the data items are in a very high dimensional space and collecting data is expensive, S_w is singular because the value for n must be kept relatively small.

One way to make classical discriminant analysis applicable to the data matrix $A \in \mathbb{R}^{m \times n}$ with $m > n$ (and hence S_w singular) is to perform dimension reduction in two stages. The discriminant analysis stage is preceded by a stage in which the cluster structure is ignored. The most popular method for the first part of this process is rank reduction by the SVD, the main tool in latent semantic indexing (LSI) [DDF+90, BDO95]. In fact, this idea has recently been implemented by Torkkola [Tor01]. However, the overall performance of this two-stage approach will be sensitive to the reduced dimension in its first stage. LSI has no theoretical optimal reduced dimension, and its computational estimation is difficult without the potentially expensive process of trying many test cases.

In this section, we extend discriminant analysis in a way that provides the optimal reduced dimension theoretically, without introducing another stage as described above. For the set of criteria involving $\text{trace}((S_2^Y)^{-1} S_1^Y)$, where S_1 and S_2 are chosen from S_w, S_b, and S_m, we use the generalized singular value decomposition (GSVD) [vL76, PS81, GV96] to extend the applicability to the case when S_2 is singular. We also establish the equivalence among alternative choices for S_1 and S_2. In Section 1.5, we address the optimization of the trace of an individual scatter matrix, and show that it can be achieved efficiently by the method of the previous section, which was derived independently of trace optimization.

1.4.1   *Generalized Singular Value Decomposition*

After the GSVD was originally defined by Van Loan [vL76], Paige and Saunders [PS81] developed the following formulation for any two matrices with the same number of columns.

THEOREM 1.2 *Suppose two matrices $K_A \in \mathbb{R}^{n \times m}$ and $K_B \in \mathbb{R}^{p \times m}$ are given. Then for*

$$K = \begin{pmatrix} K_A \\ K_B \end{pmatrix} \quad and \quad t = rank(K),$$

there exist orthogonal matrices $U \in \mathbb{R}^{n \times n}$, $V \in \mathbb{R}^{p \times p}$, $W \in \mathbb{R}^{t \times t}$, and $Q \in \mathbb{R}^{m \times m}$ such that

$$U^T K_A Q = \Sigma_A (\underbrace{W^T R,}_{t} \underbrace{0}_{m-t}) \quad and \quad V^T K_B Q = \Sigma_B (\underbrace{W^T R,}_{t} \underbrace{0}_{m-t}),$$

where

$$\underset{n \times t}{\Sigma_A} = \begin{pmatrix} I_A & & \\ & D_A & \\ & & O_A \end{pmatrix}, \quad \underset{p \times t}{\Sigma_B} = \begin{pmatrix} O_B & & \\ & D_B & \\ & & I_B \end{pmatrix},$$

and $R \in \mathbb{R}^{t \times t}$ is nonsingular with its singular values equal to the nonzero singular values of K. The matrices

$$I_A \in \mathbb{R}^{r \times r} \quad and \quad I_B \in \mathbb{R}^{(t-r-s) \times (t-r-s)}$$

are identity matrices, where

$$r = rank \begin{pmatrix} K_A \\ K_B \end{pmatrix} - rank(K_B) \quad and \quad s = rank(K_A) + rank(K_B) - rank \begin{pmatrix} K_A \\ K_B \end{pmatrix},$$

$$O_A \in \mathbb{R}^{(n-r-s) \times (t-r-s)} \quad and \quad O_B \in \mathbb{R}^{(p-t+r) \times r}$$

are zero matrices with possibly no rows or no columns, and

$$D_A = diag(\alpha_{r+1}, \ldots, \alpha_{r+s}) \quad and \quad D_B = diag(\beta_{r+1}, \ldots, \beta_{r+s})$$

satisfy

$$1 > \alpha_{r+1} \geq \cdots \geq \alpha_{r+s} > 0, \quad 0 < \beta_{r+1} \leq \cdots \leq \beta_{r+s} < 1, \qquad (1.8)$$

and $\alpha_i^2 + \beta_i^2 = 1$ for $i = r+1, \ldots, r+s$.

This form of GSVD is related to that of Van Loan by writing [PS81]

$$U^T K_A X = (\Sigma_A, 0) \quad and \quad V^T K_B X = (\Sigma_B, 0), \qquad (1.9)$$

where

$$\underset{m \times m}{X} = Q \begin{pmatrix} R^{-1}W & 0 \\ 0 & I \end{pmatrix}.$$

From the form in Eq. (1.9) we see that

$$K_A = U(\Sigma_A, 0)X^{-1} \quad \text{and} \quad K_B = V(\Sigma_B, 0)X^{-1},$$

which imply that

$$K_A^T K_A = X^{-T} \begin{pmatrix} \Sigma_A^T \Sigma_A & 0 \\ 0 & 0 \end{pmatrix} X^{-1} \quad \text{and} \quad K_B^T K_B = X^{-T} \begin{pmatrix} \Sigma_B^T \Sigma_B & 0 \\ 0 & 0 \end{pmatrix} X^{-1}.$$

Defining

$$\alpha_i = 1, \ \beta_i = 0 \text{ for } i = 1, \ldots, r$$

and

$$\alpha_i = 0, \ \beta_i = 1 \text{ for } i = r + s + 1, \ldots, t,$$

we have, for $1 \le i \le t$,

$$\beta_i^2 K_A^T K_A x_i = \alpha_i^2 K_B^T K_B x_i, \tag{1.10}$$

where x_i represents the ith column of X. For the remaining $m - t$ columns of X, both $K_A^T K_A x_i$ and $K_B^T K_B x_i$ are zero, so Eq. (1.10) is satisfied for arbitrary values of α_i and β_i when $t + 1 \le i \le m$. The columns of X are the generalized right singular vectors for the matrix pair (K_A, K_B). In terms of the generalized singular values, or the α_i / β_i quotients, r of them are infinite, s are finite and nonzero, and $t - r - s$ are zero.

1.4.2 Extension of Discriminant Analysis

For now, we focus our discussion on one of the most commonly used criteria in discriminant analysis, that of optimizing

$$J_1(G) = \text{trace}((G^T S_2 G)^{-1}(G^T S_1 G)),$$

where S_1 and S_2 are chosen from S_w, S_b, and S_m. When S_2 is assumed to be nonsingular, it is symmetric positive definite. According to results from the symmetric-definite generalized eigenvalue problem [GV96], there exists a nonsingular matrix $X \in \mathbb{R}^{m \times m}$ such that

$$X^T S_1 X = \Lambda = \text{diag}(\lambda_1 \ldots \lambda_m) \quad \text{and} \quad X^T S_2 X = I_m.$$

Since S_1 is positive semidefinite and $x_i^T S_1 x_i = \lambda_i$, each λ_i is nonnegative and only the largest $q = \text{rank}(S_1)$ λ_i's are nonzero. In addition, by using a permutation matrix to order Λ (and likewise X), we can assume that

$$\lambda_1 \ge \cdots \ge \lambda_q > \lambda_{q+1} = \cdots = \lambda_m = 0.$$

Letting x_i denote the ith column of X, we have

$$S_1 x_i = \lambda_i S_2 x_i, \tag{1.11}$$

which means that λ_i and x_i are an eigenvalue–eigenvector pair of $S_2^{-1}S_1$. We have

$$
\begin{aligned}
J_1(G) &= \text{trace}((G^T S_2 G)^{-1} G^T S_1 G) \\
&= \text{trace}((G^T X^{-T} X^{-1} G)^{-1} G^T X^{-T} \Lambda X^{-1} G) \\
&= \text{trace}((\tilde{G}^T \tilde{G})^{-1} \tilde{G}^T \Lambda \tilde{G}),
\end{aligned}
$$

where $\tilde{G} = X^{-1}G$. The matrix $\tilde{G}$ has full column rank provided G does, so it has the reduced QR factorization $\tilde{G} = QR$, where $Q \in \mathbb{R}^{m \times l}$ has orthonormal columns and R is nonsingular. Hence

$$
\begin{aligned}
J_1(G) &= \text{trace}((R^T R)^{-1} R^T Q^T \Lambda Q R) \\
&= \text{trace}(R^{-1} Q^T \Lambda Q R) \\
&= \text{trace}(Q^T \Lambda Q R R^{-1}) \\
&= \text{trace}(Q^T \Lambda Q).
\end{aligned}
$$

This shows that once we have simultaneously diagonalized S_1 and S_2, the maximization of $J_1(G)$ depends only on an orthonormal basis for $\text{range}(X^{-1}G)$; that is,

$$
\max_G J_1(G) = \max_{Q^T Q = I} \text{trace}(Q^T \Lambda Q) \le \lambda_1 + \cdots + \lambda_q = \text{trace}(S_2^{-1} S_1).
$$

(Here we consider only maximization. However, J_1 may need to be minimized for some other choices of S_1 and S_2.) When the reduced dimension $l \ge q$, this upper bound on $J_1(G)$ is achieved for

$$
Q = \begin{pmatrix} I_l \\ 0 \end{pmatrix} \quad \text{or} \quad G = X \begin{pmatrix} I_l \\ 0 \end{pmatrix} R.
$$

Note that the transformation G is not unique. That is, J_1 satisfies the invariance property $J_1(G) = J_1(GW)$ for any nonsingular matrix $W \in \mathbb{R}^{l \times l}$, since

$$
\begin{aligned}
J_1(GW) &= \text{trace}((W^T G^T S_2 GW)^{-1}(W^T G^T S_1 GW)) \\
&= \text{trace}(W^{-1}(G^T S_2 G)^{-1} W^{-T} W^T (G^T S_1 G)W) \\
&= \text{trace}((G^T S_2 G)^{-1}(G^T S_1 G)W W^{-1}) \\
&= J_1(G).
\end{aligned}
$$

Hence, the maximum $J_1(G)$ is also achieved for

$$
G = X \begin{pmatrix} I_l \\ 0 \end{pmatrix}.
$$

This means that

$$
\text{trace}((G^T S_2 G)^{-1} G^T S_1 G) = \text{trace}(S_2^{-1} S_1) \tag{1.12}
$$

whenever $G \in \mathbb{R}^{m \times l}$ consists of l eigenvectors of $S_2^{-1} S_1$ corresponding to the l largest eigenvalues.

Now, a limitation of the J_1 criteria in many applications, including text processing in information retrieval, is that the matrix S_2 must be nonsingular. Recalling the partitioning of A into k clusters given in (1.5), we define the $m \times n$ matrices

$$H_w = [A_1 - c^{(1)}e^{(1)^T}, A_2 - c^{(2)}e^{(2)^T}, \ldots, A_k - c^{(k)}e^{(k)^T}] \quad (1.13)$$

$$H_b = [(c^{(1)} - c)e^{(1)^T}, (c^{(2)} - c)e^{(2)^T}, \ldots, (c^{(k)} - c)e^{(k)^T}] \quad (1.14)$$

$$H_m = [a_1 - c, \ldots, a_n - c] = A - ce^T, \quad (1.15)$$

where $e^{(i)} = (1, \ldots, 1)^T \in \mathbb{R}^{n_i \times 1}$. Then the scatter matrices can be expressed as

$$S_w = H_w H_w^T, \ S_b = H_b H_b^T, \text{ and } S_m = H_m H_m^T. \quad (1.16)$$

For S_2 to be nonsingular, we can only allow the case $m \leq n$, since S_2 is the product of an $m \times n$ matrix and an $n \times m$ matrix [Ort87]. We seek a solution that does not impose this restriction, and which can be found without explicitly forming S_1 and S_2 from H_w, H_b, and H_m. Toward that end, we express λ_i as α_i^2 / β_i^2, and the problem (1.11) becomes

$$\beta_i^2 S_1 x_i = \alpha_i^2 S_2 x_i. \quad (1.17)$$

This has the form of a problem that can be solved using the GSVD, as described in Section 1.4.1.

We first consider the case where

$$(S_1, S_2) = (S_b, S_w).$$

From Eq. (1.16) and the definition of H_b given in Eq. (1.14), $\text{rank}(S_b) \leq k - 1$. To approximate G that satisfies both

$$\max_G \text{trace}(G^T S_b G) \quad \text{and} \quad \min_G \text{trace}(G^T S_w G), \quad (1.18)$$

we choose the x_is that correspond to the $k - 1$ largest λ_is, where $\lambda_i = \alpha_i^2 / \beta_i^2$. When the GSVD construction orders the singular value pairs as in Eq. (1.8), the generalized singular values, or the α_i / β_i quotients, are in nonincreasing order. Therefore, the first $k - 1$ columns of X are all we need. Our algorithm first computes the matrices H_b and H_w from the data matrix A. We then solve for a very limited portion of the GSVD of the matrix pair (H_b^T, H_w^T). This solution is accomplished by following the construction in the proof of Theorem 1.2 [PS81]. The major steps are limited to the complete orthogonal decomposition [GV96, LH95] of

$$K = \begin{pmatrix} H_b^T \\ H_w^T \end{pmatrix},$$

which produces orthogonal matrices P and Q and a nonsingular matrix R, followed by the singular value decomposition of a leading principal submatrix of P. The steps for this case are summarized in Algorithm DiscGSVD, adapted from [HJP03].

When $m > n$, the scatter matrix S_w is singular. Hence, we cannot even define the J_1 criterion, and discriminant analysis fails. Consider a generalized right singular vector x_i that lies in the null space of S_w. From Eq. (1.17), we see that either x_i also lies in the null space of S_b, or the corresponding β_i equals zero. We discuss each of these cases separately.

When $x_i \in \text{null}(S_w) \cap \text{null}(S_b)$, Eq. (1.17) is satisfied for arbitrary values of α_i and β_i. As explained in Section 1.4.1, this will be the case for the rightmost $m - t$ columns of X. To determine whether these columns should be included in G, consider

$$\text{trace}(G^T S_b G) = \sum g_j^T S_b g_j \quad \text{and} \quad \text{trace}(G^T S_w G) = \sum g_j^T S_w g_j,$$

where g_j represents the jth column of G. Since $x_i^T S_w x_i = 0$ and $x_i^T S_b x_i = 0$, adding the column x_i to G does not contribute to either maximization or minimization in (1.18). For this reason, we do not include these columns of X in our solution.

When $x_i \in \text{null}(S_w) - \text{null}(S_b)$, then $\beta_i = 0$. As discussed in Section 1.4.1, this implies that $\alpha_i = 1$, and hence that the generalized singular value α_i / β_i is infinite. The leftmost columns of X will correspond to these. Including these columns in G increases $\text{trace}(G^T S_b G)$, while leaving $\text{trace}(G^T S_w G)$ unchanged. We conclude that, even when S_w is singular, the rule regarding which columns of X to include in G remains the same as for the nonsingular case. The experiments in Section 1.6 demonstrate that Algorithm DiscGSVD works very well even when S_w is singular, thus extending its applicability beyond that of classical discriminant analysis.

1.4.3   Equivalence for Various S_1 and S_2

For the case when

$$(S_1, S_2) = (S_m, S_w),$$

if we follow the analysis at the beginning of Section 1.4.2 literally, it appears that we would have to include $\text{rank}(S_m) > k - 1$ columns of X in G. However, using the relation (1.7), the generalized eigenvalue problem $S_m x_i = \lambda_i S_w x_i$ can be rewritten as

$$S_b x_i = (\lambda_i - 1) S_w x_i, \quad \text{where } \lambda_i \geq 1, \ 1 \leq i \leq m.$$

In this case, the eigenvector matrix is the same as for the case of $(S_1, S_2) = (S_b, S_w)$, but the eigenvalue matrix is $\Lambda - I$. Since the same permutation can be used to put $\Lambda - I$ in nonincreasing order as was used for Λ, x_i corresponds to the ith largest eigenvalue of $S_w^{-1} S_b$. Therefore, when S_w is nonsingular, the solution is the same as for $(S_1, S_2) = (S_b, S_w)$.

When $m > n$, the scatter matrix S_w is singular. For a generalized right singular vector $x_i \in \text{null}(S_w)$, $S_m x_i = S_b x_i$. Hence, we include the same columns in G as we did for $(S_1, S_2) = (S_b, S_w)$. Alternatively, we can show that the solutions are the same by deriving a GSVD of the matrix pair (H_m^T, H_w^T) that has the same generalized right singular vectors as (H_b^T, H_w^T). See [HP02] for the details.

Note that in the m-dimensional space,

$$\text{trace}(S_w^{-1} S_m) = \text{trace}(S_w^{-1}(S_w + S_b)) = m + \text{trace}(S_w^{-1} S_b), \qquad (1.19)$$

Algorithm 1.3 DiscGSVD

Given a data matrix $A \in \mathbb{R}^{m \times n}$ with k clusters, compute the columns of the matrix $G \in \mathbb{R}^{m \times (k-1)}$, which preserves the cluster structure in the reduced dimensional space, using

$$J_1(G) = \text{trace}((G^T S_w G)^{-1} G^T S_b G).$$

Also compute the $k - 1$ dimensional representation Y of A.

1. Compute H_b and H_w from A according to

    $$H_b = [\sqrt{n_1}(c^{(1)} - c), \sqrt{n_2}(c^{(2)} - c), \ldots, \sqrt{n_k}(c^{(k)} - c)] \in \mathbb{R}^{m \times k},$$

    and (1.13), respectively. (Using this equivalent but lower-dimensional form of H_b reduces complexity.)

2. Compute the complete orthogonal decomposition of

    $$K = \begin{pmatrix} H_b^T \\ H_w^T \end{pmatrix} \in \mathbb{R}^{(k+n) \times m}, \text{ which is } P^T K Q = \begin{pmatrix} R & 0 \\ 0 & 0 \end{pmatrix}.$$

3. Let $t = \text{rank}(K)$.

4. Compute W from the SVD of $P(1 : k, 1 : t)$, which is
    $U^T P(1 : k, 1 : t) W = \Sigma_A.$

5. Compute the first $k - 1$ columns of $X = Q \begin{pmatrix} R^{-1} W & 0 \\ 0 & I \end{pmatrix}$,

    and assign them to G.

6. $Y = G^T A$.

---

and in the l-dimensional space,

$$\text{trace}((S_w^Y)^{-1} S_m^Y) = \text{trace}((S_w^Y)^{-1}(S_w^Y + S_b^Y)) = l + \text{trace}((S_w^Y)^{-1} S_b^Y). \quad (1.20)$$

This confirms that the solutions are the same for both $(S_1, S_2) = (S_b, S_w)$ and $(S_1, S_2) = (S_m, S_w)$. For any $l \geq k - 1$, when G includes the eigenvectors of $S_w^{-1} S_b$ corresponding to the l largest eigenvalues, then

$$\text{trace}(S_w^{-1} S_b) = \text{trace}((S_w^Y)^{-1} S_b^Y).$$

By subtracting (1.20) from (1.19), we get

$$\text{trace}((S_w^Y)^{-1} S_m^Y) + (m - l) = \text{trace}(S_w^{-1} S_m). \quad (1.21)$$

In other words, each additional eigenvector beyond the leftmost $k - 1$ will add one to $\text{trace}((S_w^Y)^{-1} S_m^Y)$. This shows that we do not preserve the cluster structure when measured by $\text{trace}(S_w^{-1} S_m)$, although we do preserve $\text{trace}(S_w^{-1} S_b)$. According to Eq. (1.21), $\text{trace}(S_w^{-1} S_m)$ will be preserved if we include all $\text{rank}(S_m) = m$ eigenvectors of $S_w^{-1} S_m$.

For the case

$$(S_1, S_2) = (S_w, S_m),$$

we want to minimize trace$(S_m^{-1} S_w)$. In [HP02], we use a similar argument to show that the solution is the same as for $(S_1, S_2) = (S_b, S_w)$, even when S_w is singular. However, since we are minimizing in this case, the generalized singular values are in nondecreasing order, taking on reciprocal values of those for (H_m^T, H_w^T).

Having shown the equivalence of the J_1 criteria for various (S_1, S_2), we conclude that $(S_1, S_2) = (S_b, S_w)$ should be used for the sake of computational efficiency. The DiscGSVD algorithm reduces computational complexity further by using a lower-dimensional form of H_b rather than that presented in Eq. (1.14), and it avoids a potential loss of information [GV96, page 239, Example 5.3.2] by not explicitly forming S_b and S_w as cross-products of H_b and H_w.

1.5   Trace Optimization Using an Orthogonal Basis of Centroids

Simpler criteria for preserving cluster structure, such as min trace$(G^T S_w G)$ and max trace$(G^T S_b G)$, involve only one of the scatter matrices. A straightforward minimization of trace$(G^T S_w G)$ seems meaningless since the optimum always reduces the dimension to one, even when the solution is restricted to the case when G has orthonormal columns. On the other hand, with the same restriction, maximization of trace$(G^T S_b G)$ produces an equivalent solution to the CentroidQR method, which was discussed in Section 1.3.

Let

$$J_2(G) = \text{trace}(G^T S_b G).$$

If we let $G \in \mathbb{R}^{m \times l}$ be any matrix with full column rank, then essentially there is no upper bound and maximization is also meaningless. Now let us restrict the solution to the case when G has orthonormal columns. Then there exists $\hat{G} \in \mathbb{R}^{m \times (m-l)}$ such that $(G, \hat{G})$ is an orthogonal matrix. In addition, since S_b is positive semidefinite, we have

$$\text{trace}(G^T S_b G) \leq \text{trace}(G^T S_b G) + \text{trace}(\hat{G}^T S_b \hat{G}) = \text{trace}(S_b).$$

If the SVD of H_b is given by $H_b = U \Sigma V^T$, then $S_b U = U \Sigma \Sigma^T$. Hence the columns of U form an orthonormal set of eigenvectors of S_b corresponding to the nonincreasing eigenvalues on the diagonal of $\Lambda = \Sigma \Sigma^T$. For $p = \text{rank}(S_b)$, if we let U_p denote the first p columns of U and $\Lambda_p = \text{diag}(\lambda_1 \ldots \lambda_p)$, we have

$$\begin{aligned}
J_2(U_p) &= \text{trace}(U_p^T S_b U_p) \\
&= \text{trace}(U_p^T U_p \Lambda_p) \\
&= \lambda_1 + \cdots + \lambda_p \\
&= \text{trace}(S_b).
\end{aligned}$$

This means that we preserve trace(S_b) if we take U_p as G.

Now we show that this solution is equivalent to the solution of the CentroidQR method, which does not involve the computation of eigenvectors. Defining the centroid matrix $C = [c^{(1)} \cdots c^{(k)}]$ as in Algorithm 1.1, C has the reduced QR decomposition $C = Q_k R$, where $Q_k \in \mathbb{R}^{m \times k}$ has orthonormal columns and $R \in \mathbb{R}^{k \times k}$ [?]. Suppose x is an eigenvector of S_b corresponding to the nonzero eigenvalue λ. Then

$$S_b x = \sum_{i=1}^{k} n_i (c^{(i)} - c)(c^{(i)} - c)^T x = \lambda x.$$

This means $x \in \text{span}\{c^{(i)} - c | 1 \le i \le k\}$, and hence $x \in \text{span}\{c^{(i)} | 1 \le i \le k\}$. Accordingly,

$$\text{range}(U_p) \subseteq \text{range}(C) \subseteq \text{range}(Q_k),$$

which implies that $U_p = Q_k W$ for some matrix $W \in \mathbb{R}^{k \times p}$ with orthonormal columns. This yields

$$
\begin{aligned}
J_2(U_p) &= \text{trace}(W^T Q_k^T S_b Q_k W) \\
&\le \text{trace}(Q_k^T S_b Q_k) \\
&= J_2(Q_k).
\end{aligned}
$$

Hence

$$J_2(Q_k) = \text{trace}(S_b),$$

and Q_k plays the same role as U_p. In other words, instead of computing the eigenvectors, we simply need to compute Q_k, which is much cheaper. Therefore, by computing a reduced QR decomposition of the centroid matrix, we obtain a solution that maximizes $\text{trace}(G^T S_b G)$ over all G with orthonormal columns.

1.6   Document Classification Experiments

In this section, we demonstrate the effectiveness of the DiscGSVD and CentroidQR algorithms, which use the J_1 criterion with $(S_1, S_2) = (S_b, S_w)$ and the J_2 criterion with $G^T G = I$, respectively. For DiscGSVD, we confirm its mathematical equivalence to J_1 using an alternative choice of (S_1, S_2), and we illustrate the discriminatory power of J_1 via two-dimensional projections. Just as important, we validate our extension of J_1 to the singular case. For CentroidQR, its preservation of $\text{trace}(S_b)$ is shown to be a very effective compromise for the simultaneous optimization of two traces approximated by J_1.

In Table 1.1, we use clustered data that are artificially generated by an algorithm adapted from [JD88, Appendix H]. The data consist of 2000 documents in a space of dimension 150, with $k = 7$ clusters. DiscGSVD reduces the dimension from 150 to $k - 1 = 6$. We compare the DiscGSVD criterion, $J_1 = \text{trace}(S_w^{-1} S_b)$, with the alternative J_1 criterion, $\text{trace}(S_w^{-1} S_m)$. The trace values confirm our theoretical

Method	Full	Trace($S_w^{-1}S_b$)	Trace($S_w^{-1}S_m$)	
Dim	150×2000	6×2000	6×2000	7×2000
trace(S_w)	299700	1.97	1.48	1.98
trace(S_b)	22925	4.03	3.04	3.04
trace(S_m)	322630	6.00	4.52	5.02
trace($S_w^{-1}S_b$)	12.6	12.6	12.6	12.6
trace($S_w^{-1}S_m$)	162.6	18.6	18.6	19.6
centroid	2.6 %	2.2 %	2.0 %	2.0 %
5nn	18.7 %	2.2 %	2.2 %	2.4 %
15nn	10.1 %	1.8 %	1.9 %	2.1 %

Table 1.1. Traces and Misclassification Rates (in %) with L_2 Norm Similarity

---

Algorithm 1.4 Centroid-Based Classification

Given a data matrix A with k clusters and k corresponding centroids, $c^{(i)}$ for $1 \leq i \leq k$, find the index j of the cluster to which a vector q belongs.

- Find the index j such that $sim(q, c^{(i)})$, $1 \leq i \leq k$, is minimum (or maximum), where $sim(q, c^{(i)})$ is the similarity measure between q and $c^{(i)}$.

(For example, $sim(q, c^{(i)}) = \|q - c^{(i)}\|_2$ using the L_2 norm, and we take the index with the minimum value. Using the cosine measure, $sim(q, c^{(i)}) = cos(q, c^{(i)}) = \frac{q^T c^{(i)}}{\|q\|_2 \|c^{(i)}\|_2}$, and we take the index with the maximum value.)

---

findings, namely, that the generalized eigenvectors that optimize the alternative J_1 also optimize DiscGSVD's J_1, and including an additional eigenvector increases trace($S_w^{-1}S_m$) by one.

We also report misclassification rates for a centroid-based classification method [HJP03] and the k-nearest neighbor (knn) classification method [TK99], which are summarized in Algorithms 1.4 and 1.5. (Note that the classification parameter of knn differs from the number of clusters k.) These are obtained using the L_2 norm or Euclidean distance similarity measure. While these rates differ slightly with the choice of S_b or S_m, and the reduction to six or seven rows using the latter, they establish no advantage of using S_m over S_b, even when we include an additional eigenvector to bring us closer to the preservation of trace($S_w^{-1}S_m$). These results bolster our argument that the correct choice of J_1 is optimized in our DiscGSVD algorithm, since it limits the GSVD computation to a composite matrix with $k + n$ rows, rather than one with $2n$ rows.

To illustrate the power of the J_1 criterion, we use it to reduce the dimension from 150 to two. Even though the optimal reduced dimension is six, J_1 does surprisingly well at discriminating among seven classes, as seen in Figure 1.1. As expected,

---

Algorithm 1.5 k Nearest Neighbor (knn) Classification

Given a data matrix $A = [a_1 \quad \cdots \quad a_n]$ with k clusters, find the cluster to which a vector q belongs.

1. From the similarity measure $sim(q, a_j)$ for $1 \leq j \leq n$, find the k nearest neighbors of q. (We use k to distinguish the algorithm parameter from the number of clusters k.)

2. Among these k vectors, count the number belonging to each cluster.

3. Assign q to the cluster with the greatest count in the previous step.

---

the alternative J_1 does equally well in Figure 1.2. In contrast, Figure 1.3 shows that the truncated SVD is not the best discriminator.

Another set of experiments validates our extension of J_1 to the singular case. For this purpose, we use five categories of abstracts from the MEDLINE [1] database (see Table 1.2). Each category has 40 documents. There are 7519 terms after pre-processing with stemming and removal of stopwords [Kow97]. Since 7519 exceeds the number of documents (200), S_w is singular and classical discriminant analysis breaks down. However, our DiscGSVD method circumvents this singularity problem.

The DiscGSVD algorithm dramatically reduces the dimension 7519 to four, or one less than the number of clusters. The CentroidQR method reduces the dimension to five. Table 1.3 shows classification results using the L_2 norm similarity measure. DiscGSVD produces the lowest misclassification rate using both centroid-based and nearest neighbor classification methods. Because the J_1 criterion is not defined in this case, we compute the ratio trace(S_b)/trace(S_w) as a rough optimality measure. We observe that the ratio is strikingly higher for DiscGSVD reduction than for the other methods. These experimental results confirm that the DiscGSVD algorithm effectively extends the applicability of the J_1 criterion to cases that classical discriminant analysis cannot handle. In addition, the CentroidQR algorithm preserves trace(S_b) from the full dimension without the expense of computing eigenvectors. Taken together, the results for these two methods demonstrate the potential for dramatic and efficient dimension reduction without compromising cluster structure.

1.7   Conclusion

Our experimental results verify that the J_1 criterion, when applicable, effectively optimizes classification in the reduced dimensional space, while our DiscGSVD extends the applicability to cases that classical discriminant analysis cannot handle.

---

[1] http://www.ncbi.nlm.nih.gov/PubMed.

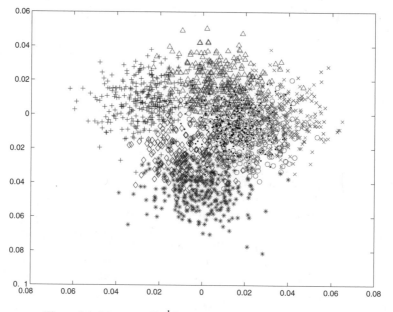

Figure 1.1. Max trace$(S_w^{-1} S_b)$ projection onto two dimensions.

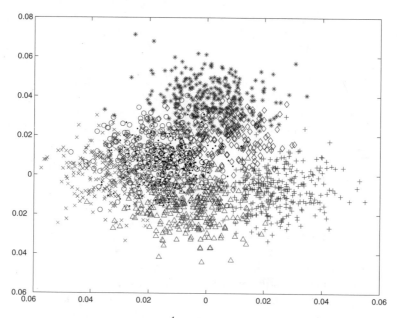

Figure 1.2. Max trace$(S_w^{-1} S_m)$ projection onto two dimensions.

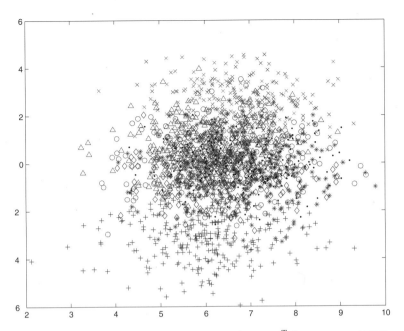

Figure 1.3. Two-dimensional representation using $\Sigma_2 V_2^T$ from truncated SVD.

Class	Category	No. of Documents
1	heart attack	40
2	colon cancer	40
3	diabetes	40
4	oral cancer	40
5	tooth decay	40
	dimension	7519×200

Table 1.2. MEDLINE Data Set

Method		Full	CentroidQR	DiscGSVD
Dim		7519×200	5×200	4×200
trace values	$\mathrm{trace}(S_w)$	73048	4210	0.05
	$\mathrm{trace}(S_b)$	6229	6229	3.95
	$\frac{\mathrm{trace}(S_b)}{\mathrm{trace}(S_w)}$	0.09	1.5	79
misclassification rate in %	centroid	5	5	1
	1nn	40	3	1

Table 1.3. Traces and Misclassification Rate with L_2 Norm Similarity

In addition, our DiscGSVD algorithm avoids the numerical problems inherent in explicitly forming the scatter matrices.

In terms of computational complexity, the most expensive part of Algorithm DiscGSVD is Step 2, where a complete orthogonal decomposition is needed. Assuming $k \le n$, $t \le m$, and $t = \mathcal{O}(n)$, the complete orthogonal decomposition of K costs $\mathcal{O}(nmt)$ when $m \le n$, and $\mathcal{O}(m^2 t)$ when $m > n$ [GV96]. Therefore, a fast algorithm needs to be developed for Step 2.

For CentroidQR, the most expensive step is the reduced QR decomposition of C, which costs $\mathcal{O}(mk^2)$ [GV96]. By solving a simpler eigenvalue problem and avoiding the computation of eigenvectors, CentroidQR is significantly cheaper than DiscGSVD. Our experiments show it to be a very reasonable compromise.

Finally, it bears repeating that dimension reduction is only a preprocessing stage. Since classification and document retrieval will be the dominating parts computationally, the expense of dimension reduction should be weighed against its effectiveness in reducing the cost involved in those processes.

Acknowledgments: This work was supported in part by the National Science Foundation grants CCR-9901992 and CCR-0204109.

A part of this work was carried out while H. Park was visiting the Korea Institute for Advanced Study in Seoul, Korea, for her sabbatical leave, from September 2001 to August 2002.

References

[BDO95]   M. Berry, S. Dumais, and G. O'Brien.Using linear algebra for intelligent information retrieval.*SIAM Review*, 37(4):573–595, 1995.

[Bjö96]   Å. Björck.*Numerical Methods for Least Squares Problems*.SIAM, Philadelphia, 1996.

[CF79]   R.E. Cline and R.E. Funderlic.The rank of a difference of matrices and associated generalized inverses.*Linear Algebra Appl.*, 24:185–215, 1979.

[CFG95]   M.T. Chu, R.E. Funderlic, and G.H. Golub.A rank-one reduction formula and its applications to matrix factorizations.*SIAM Review*, 37(4):512–530, 1995.

[DDF+90]   S. Deerwester, S. Dumais, G. Furnas, T. Landauer, and R. Harshman.Indexing by latent semantic analysis.*Journal of the American Society for Information Science*, 41(6):391–407, 1990.

[DHS01]   R.O. Duda, P.E. Hart, and D.G. Stork.*Pattern Classification*, second edition.Wiley, New York, 2001.

[Fuk90]   K. Fukunaga.*Introduction to Statistical Pattern Recognition*, second edition.Academic, Boston, MA, 1990.

[GV96]   G. Golub and C. Van Loan.*Matrix Computations*, third edition.John Hopkins Univ. Press, Baltimore, MD, 1996.

[Gut57]   L. Guttman.A necessary and sufficient formula for matric factoring.*Psychometrika*, 22(1):79–81, 1957.

[HJP03]   P. Howland, M. Jeon, and H. Park.Structure preserving dimension reduction for clustered text data based on the generalized singular value decomposition.*SIAM Journal on Matrix Analysis and Applications*, 2003, to appear.

[HMH00]   L. Hubert, J. Meulman, and W. Heiser.Two purposes for matrix factorization: A historical appraisal.*SIAM Review*, 42(1):68–82, 2000.

[Hor65]   P. Horst.*Factor Analysis of Data Matrices*.Holt, Rinehart and Winston, Orlando, FL, 1965.

[HP02]   P. Howland and H. Park.Extension of discriminant analysis based on the generalized singular value decomposition.Technical Report 021, Department of Computer Science and Engineering, University of Minnesota, Twin Cities, 2002.

[JD88]   A. Jain and R. Dubes.*Algorithms for Clustering Data*.Prentice-Hall, Englewood Cliffs, NJ, 1988.

[Kow97]   G. Kowalski.*Information Retrieval Systems: Theory and Implementation*. Kluwer Academic, Hingham, MA, 1997.

[LH95]   C.L. Lawson and R.J. Hanson.*Solving Least Squares Problems*.SIAM, Philadelphia, 1995.

[Ort87]   J. Ortega.*Matrix Theory: A Second Course*.Plenum, New York, 1987.

[PJR03]   H. Park, M. Jeon, and J.B. Rosen.Lower dimensional representation of text data based on centroids and least squares.*BIT*, 2003, to appear.

[PS81]   C.C. Paige and M.A. Saunders.Towards a generalized singular value decomposition.*SIAM Journal on Numerical Analysis*, 18(3):398–405, 1981.

[Sal71]   G. Salton.*The SMART Retrieval System*.Prentice-Hall, Englewood Cliffs, NJ, 1971.

[SM83]   G. Salton and M.J. McGill.*Introduction to Modern Information Retrieval*.McGraw-Hill, New York, 1983.

[Thu45]   L.L. Thurstone.A multiple group method of factoring the correlation matrix.*Psychometrika*, 10(2):73–78, 1945.

[TK99]   S. Theodoridis and K. Koutroumbas.*Pattern Recognition*.Academic, San Diego, 1999.

[Tor01]   K. Torkkola.Linear discriminant analysis in document classification.In *Proceedings of the IEEE ICDM Workshop on Text Mining*, 2001.

[vL76]   C.F. Van Loan.Generalizing the singular value decomposition.*SIAM Journal on Numerical Analysis*, 13(1):76–83, 1976.

2

Automatic Discovery of Similar Words

Pierre P. Senellart
Vincent D. Blondel

Overview

We deal with the issue of automatic discovery of similar words (synonyms and near-synonyms) from different kinds of sources: from large corpora of documents, from the Web, and from monolingual dictionaries. We present in detail three algorithms that extract similar words from a large corpus of documents and consider the specific case of the World Wide Web. We then describe a recent method of automatic synonym extraction in a monolingual dictionary. The method is based on an algorithm that computes similarity measures between vertices in graphs. We use the 1913 *Webster's Dictionary* and apply the method on four synonym queries. The results obtained are analyzed and compared with those obtained by two other methods.

2.1 Introduction

The purpose of this chapter is to review some methods used for automatic extraction of similar words from different kinds of sources: large corpora of documents, the Web, and monolingual dictionaries. The underlying goal of these methods is the automatic discovery of synonyms. This goal is, in general, too difficult to achieve since it is often difficult to distinguish in an automatic way synonyms, antonyms, and, more generally, words that are semantically close to each other. Most methods provide words that are "similar" to each other. We mainly describe two kinds of methods: techniques that, upon input of a word, automatically compile a list of good synonyms or near-synonyms, and techniques that generate a thesaurus (from some source, they build a complete lexicon of related words). They differ because in the latter case, a complete thesaurus is generated at the same time and there may not be an entry in the thesaurus for each word in the source. Nevertheless, the purposes of the techniques are very similar and we therefore do not distinguish much between them.

There are many applications of such methods. For example, in natural language processing and information retrieval they can be used to broaden and modify natural language queries. They can also be used as a support for the compilation of synonym dictionaries, which is a tremendous task. In this chapter we focus on the search for synonyms rather than on applications of these techniques.

Many approaches for the automatic construction of thesauri from large corpora have been proposed. Some of them are presented in Section 2.2. The value of such domain-specific thesauri, as opposed to general handmade synonym dictionaries is stressed. We also look at the particular case of the Web, whose large size and other specific features do not allow their being handled in the same way as more classical corpora. In Section 2.3, we propose an original approach, which is based on monolingual dictionaries and uses an algorithm that generalizes an algorithm initially proposed by Kleinberg for searching the Web. Two other methods working from monolingual dictionaries are also presented.

2.2   Discovery of Similar Words from a Large Corpus

Much research has been carried out on the search for similar words in corpora, mostly for applications in information retrieval tasks. A large number of these approaches are based on the simple assumption that similar words are used in the same contexts. The methods differ in the way the contexts are defined (the document, a textual window, or more or less elaborate syntactical contexts) and the way the similarity is computed.

Depending on the type of corpus, we may obtain different emphasis in the resulting lists of synonyms. The thesaurus built from a corpus is domain-specific to this corpus and is thus more adapted to a particular application in this domain than a general hand-written dictionary. There are several other advantages to the use of computer-written thesauri. In particular, they may be rebuilt easily to mirror a change in the collection of documents (and thus in the corresponding field), and they are not biased by the lexicon writer (but are, of course, biased by the corpus in use). Obviously, however, hand-written synonym dictionaries are bound to be more liable, with fewer gross mistakes.

We describe below three methods that may be used to discover similar words. Of course, we do not pretend to be exhaustive, but rather have chosen to present some of the main approaches. In Section 2.2.1, we present a straightforward method, involving a document vector space model and the cosine similarity measure. This method is used by Chen and Lynch to extract information from a corpus on East-bloc computing [CL92] and we briefly report their results. We then look at an approach proposed by Crouch [Cro90] for the automatic construction of a thesaurus. The method is based on a term vector space model and term discrimination values [SYY75], and is specifically adapted for words that are not too frequent. In

Section 2.2.3, we focus on Grefenstette's SEXTANT system [Gre94], which uses a partial syntactical analysis. Finally, in the last section, we consider the particular case of the Web as a corpus, and discuss the problem of finding synonyms in a very large collection of documents.

2.2.1 A Document Vector Space Model

The first obvious definition of the context, given a collection of documents, is to say that terms are similar if they tend to occur in the same documents. This can be represented in a multidimensional space, where each document is a dimension and each term is a vector in document space with Boolean entries indicating whether the term appears in the corresponding document. It is common in information retrieval to use this type of vector space model. In the dual model, terms are coordinates and documents are vectors in term space; we show an application of this dual model in the next section.

Thus two terms are similar if their corresponding vectors are close to each other. The similarity between the vector $\mathbf{i}$ and the vector $\mathbf{j}$ is computed using a similarity measure, such as the cosine:

$$\cos(\mathbf{i}, \mathbf{j}) = \frac{\mathbf{i} \cdot \mathbf{j}}{\sqrt{\mathbf{i} \cdot \mathbf{i} \times \mathbf{j} \cdot \mathbf{j}}} \, ,$$

where $\mathbf{i} \cdot \mathbf{j}$ is the inner product of $\mathbf{i}$ and $\mathbf{j}$. With this definition, we have $0 \le \cos(\mathbf{i}, \mathbf{j}) \le 1$; θ with $\cos\theta = \cos(\mathbf{i}, \mathbf{j})$ is the angle between $\mathbf{i}$ and $\mathbf{j}$. Similar terms will tend to occur in the same documents and the angle between them will be small. Thus the cosine similarity measure will be close to one. In contrast, terms with little in common will not occur in the same documents, the angle between them will be close to $\pi/2$, and the cosine similarity measure will be close to zero.

The cosine is a commonly used similarity measure. One must, however, not forget that the justification of its use is based on the assumption that the axes are orthogonal, which is seldom the case in practice since documents in the collection are bound to have something in common and not be completely independent.

In [CL92] Chen and Lynch compare the cosine measure with another measure, referred to as the Cluster measure. The Cluster measure is asymmetrical, thus giving asymmetrical similarity relationships between terms. It is defined by

$$cluster(\mathbf{i}, \mathbf{j}) = \frac{\mathbf{i} \cdot \mathbf{j}}{\|\mathbf{i}\|_1} \, ,$$

where $\|\mathbf{i}\|_1$ is the sum of the magnitudes of $\mathbf{i}$'s coordinates (i.e., the l_1 norm of $\mathbf{i}$).

For both these similarity measures the algorithm is then straightforward: once a similarity measure has been selected, its value is computed between every pair of terms, and the best similar terms are kept for each term.

The corpus Chen and Lynch worked on was a 200 MB collection of various text documents on computing in the former East-bloc countries. They did not run the algorithms on the raw text. The whole database was manually annotated so that every document was assigned a list of appropriate keywords, countries, organization names, journal names, person names, and folders. Around 60,000 terms were obtained in this way and the similarity measures were computed on them.

For instance, the best similar keywords (with the cosine measure) for the keyword **technology transfer** were: **export controls**, **trade**, **covert**, **export**, **import**, **micro-electronics**, **software**, **microcomputer**, and **microprocessor**. These are indeed related (in the context of the corpus) and words such as **trade**, **import**, and **export** are likely to be some of the best near-synonyms in this context.

The two similarity measures were compared on randomly chosen terms with lists of words given by human experts in the field. Chen and Lynch report that the Cluster algorithm presents a better Concept Recall ratio (i.e., is, the proportion of relevant terms that were selected) than cosine and human experts. Both similarity measures exhibits similar Concept Precision ratios (i.e., the proportion of selected terms that were relevant), and they are inferior to that of human experts. The asymmetry of Cluster seems to be a real advantage.

2.2.2   A Thesaurus of Infrequent Words

In [Cro90] Crouch presents a method for the automatic construction of thesaurus classes regrouping words that appear seldom in the corpus. Her purpose is to use this thesaurus to modify queries asked of an information retrieval system. She uses a term vector space model, which is the dual of the space used in the previous section: words are dimensions and documents are vectors. The projection of a vector along an axis is the weight of the corresponding word in the document. Different weighting schemes might be used; one that seems effective is the "Term Frequency Inverse Document Frequency" (*TF-IDF*), that is, the number of times the word appears in the document multiplied by a (monotone) function of the inverse of the number of documents in which the word appears. Terms that appear often in a document and do not appear in many documents therefore have an important weight.

As we saw earlier, we can use a similarity measure such as the cosine to characterize the similarity between two vectors (i.e., two documents). The algorithm proposed by Crouch, presented in more detail below, is to cluster the set of documents according to this similarity and then to select *indifferent discriminators* from the resulting clusters to build thesaurus classes.

Salton, Yang, and Yu introduce in [SYY75] the notion of *term discrimination value*. It is a measure of the effect of the addition of a term (as a dimension) to the vector space on the similarities between documents. A good discriminator is a term that tends to raise the distances between documents; a poor discriminator tends to lower the distances between documents; finally, an indifferent discriminator does not change the distances between documents much. The exact or approximate computation of all term discrimination values is an expensive task. To avoid this problem, the authors propose using the term document frequency (i.e., the number of documents the term appears in) instead of the discrimination value, since experiments show they are strongly related. Terms appearing in less than about 1% of the documents are mostly indifferent discriminators; terms appearing in more than 1% and less than 10% of the documents are good discriminators; very frequent terms are poor discriminators.

Crouch therefore suggests using low-frequency terms to form thesaurus classes, which should be made of indifferent discriminators. The first idea to build the thesaurus would be to cluster these low-frequency terms with an adequate clustering algorithm. This is not very interesting, however, since, by definition, one does not have much information about low-frequency terms. But the documents themselves may be clustered in a meaningful way. The complete link clustering algorithm, which produces small and tight clusters, is adapted to the problem. Each document is first considered as a cluster by itself, and iteratively, the two closest clusters (the similarity between clusters is defined to be the minimum of all similarities (computed by the cosine measure) between a pair of documents in the two clusters) are merged, until the distance between clusters becomes higher than a user-supplied threshold.

When this clustering step is performed, low-frequency words are extracted from each cluster. They build corresponding thesaurus classes. Crouch does not describe these classes but has used them directly for broadening information retrieval queries, and has observed substantial improvements in both recall and precision on two classical test corpora. It is therefore legitimate to assume that words in the thesaurus classes are related to each other. This method only works on low-frequency words, but the other methods presented here have problems in dealing with such words for which we have little information.

2.2.3   The SEXTANT System

Grefenstette presents in [Gre93, Gre94] an algorithm for the discovery of similar words that uses a partial syntactical analysis. The different steps of the algorithm SEXTANT (Semantic EXtraction from Text via Analyzed Networks of Terms) are detailed below.

Lexical Analysis

Words in the corpus are separated using a simple lexical analysis. A proper name analyzer is also applied. Then each word is looked up in a lexicon and is assigned a part of speech. If a word has several possible parts of speech, a disambiguator is used to choose the most probable one.

Noun and Verb Phrase Bracketing

Noun and verb phrases are then detected in the sentences of the corpus, using starting, ending, and continuation rules: for instance, a determiner can start a noun phrase, a noun can follow a determiner in a noun phrase, an adjective can not start, end, or follow any kind of word in a verb phrase, and so on.

ADJ	:	an adjective modifies a noun	(e.g., civil unrest)
NN	:	a noun modifies a noun	(e.g., animal rights)
NNPREP	:	a noun that is the object of a preposition modifies a preceding noun	(e.g., measurements along the crest)
SUBJ	:	a noun is the subject of a verb	(e.g., the table shook)
DOBJ	:	a noun is the direct object of a verb	(e.g., shook the table)
IOBJ	:	a noun in a prepositional phrase modifying a verb	(e.g., the book was placed on the table)

Figure 2.1. Syntactical relations extracted by SEXTANT.

Parsing

Several syntactic relations (or contexts) are then extracted from the bracketed sentences, requiring five successive passes over the text. Figure 2.1, taken from [Gre94], shows the list of extracted relations.

The relations generated are thus not perfect (on a sample of 60 sentences Grefenstette found a correctness ratio of 75%) and could be better if a more elaborate parser were used, but it would be more expensive too. Five passes over the text are enough to extract these relations, and since the corpus dealt with may be very large, backtracking, recursion, or other time-consuming techniques used by elaborate parsers would be inappropriate.

Similarity

Grefenstette focuses on the similarity between nouns; other parts of speech are not discussed. After the parsing step, a noun has a number of attributes: all the words that modify it, along with the kind of syntactical relation (ADJ for an adjective, NN or NNPREP for a noun, and SUBJ, DOBJ, or IOBJ for a verb). For instance, the noun **cause**, which appears 83 times in a corpus of medical abstracts, has 67 unique attributes in this corpus. These attributes constitute the context of the noun, on which similarity computations will be made. Each attribute is assigned a weight by

$$weight(att) = 1 + \sum_{\text{noun } i} \frac{p_{att,i} log(p_{att,i})}{log(\text{total number of relations})},$$

where

$$p_{att,i} = \frac{\text{number of times } att \text{ appears with } i}{\text{total number of attributes of } i}.$$

The similarity measure used by Grefenstette is a weighted Jaccard similarity measure defined as follows

1. CRAN (Aeronautics abstract)
   case: **characteristic, analysis, field, distribution, flaw, number, layer, problem**

2. JFK (Articles on JFK assassination conspiracy theories)
   case: **film, evidence, investigation, photograph, picture, conspiracy, murder**

3. MED (Medical abstracts)
   case: **change, study, patient, result, treatment, child, defect, type, disease, lesion**

Figure 2.2. SEXTANT similar words for **case**, from different corpora.

species	**bird, fish, family, group, form, animal, insect, range, snake**
fish	**animal, species, bird, form, snake, insect, group, water**
bird	**species, fish, animal, snake, insect, form, mammal, duck**
water	**sea, area, region, coast, forest, ocean, part, fish, form, lake**
egg	**nest, female, male, larva, insect, day, form, adult**

Figure 2.3. SEXTANT similar words for words with most contexts in *Grolier's Encyclopedia* animal articles.

$$jac(\mathbf{i}, \mathbf{j}) = \frac{\sum_{att \text{ attribute of both } \mathbf{i} \text{ and } \mathbf{j}} weight(att)}{\sum_{att \text{ attribute of either } \mathbf{i} \text{ or } \mathbf{j}} weight(att)} .$$

Results

Grefenstette used SEXTANT on various corpora and many examples of the results returned are available in [Gre94]. Figure 2.2 shows the most similar words of **case** in three completely different corpora. It is interesting to note that the corpus has a great impact on the meaning of the word according to which similar words are selected. This is a good illustration of the value of working on a domain-specific corpus.

Figure 2.3 shows other examples, in a corpus on animals. Most words are closely related to the initial word and some of them are indeed very good (**sea, ocean, lake** for **water**; **family, group** for species, ...). There remain completely unrelated words though, such as **day** for **egg**.

2.2.4   How to Deal with the Web

The World Wide Web is a very particular corpus: its size can simply not be compared with the largest corpora traditionally used for synonym extraction, its access times are high, and it is also richer and more lively than any other corpus. Moreover, a large part of it is conveniently indexed by search engines. One could imagine that its hyperlinked structure could be of some use too. And of course it is not a domain-specific thesaurus. Is it possible to use the Web for the discovery of similar words? Obviously, because of the size of the Web, none of the above techniques can apply.

Turney partially deals with the issue in [Tur01]. He does not try to obtain a list of synonyms of a word **i** but, given a word **i**, he proposes a way to assign a synonymy score to any word **j**. His method was checked on synonym recognition questions extracted from two English tests: the Test Of English as a Foreign Language (TOEFL) and the English as a Second Language test (ESL). Four different synonymy scores are compared. They use the advanced search functions of the Altavista search engine (http://www.altavista.com).

$$score_1(j) = \frac{hits(i \; AND \; j)}{hits(j)}$$

$$score_2(j) = \frac{hits(i \; NEAR \; j)}{hits(j)}$$

$$score_3(j) = \frac{hits((i \; NEAR \; j) \; AND \; NOT \; ((i \; OR \; j) \; NEAR \; \text{"not"}))}{hits(j \; AND \; NOT \; (j \; NEAR \; \text{"not"}))}$$

$$score_4(j) = \frac{hits((i \; NEAR \; j) \; AND \; \textbf{context} \; AND \; NOT \; ((i \; OR \; j) \; NEAR \; \text{"not"}))}{hits(j \; AND \; \textbf{context} \; AND \; NOT \; (j \; NEAR \; \text{"not"}))} \; .$$

In these expressions, $hits$ represents the number of pages returned by Altavista for the corresponding query; AND, OR, and NOT are the classical Boolean operators; $NEAR$ imposes that the two words not be separated by more than 10 words; and **context** is a context word (a context was given along with the question in ESL; the context word may be automatically derived from it). The difference between $score_2$ and $score_3$ was introduced in order not to assign good scores to antonyms.

The four scores are presented in increasing order of the quality of the corresponding results. $score_3$ gives a good synonym for 73.75% of the questions from TOEFL ($score_4$ was not applicable since no context was given) and $score_4$ gives a good synonym in 74% of the questions from ESL. These results are arguably good, since, as reported by Turney, the average score of TOEFL by a large sample of students is 64.5%.

This algorithm cannot be used to obtain a list of synonyms, since it is too expensive to run for each candidate word in a dictionary because of network access times, but it may be used, for instance, to refine a list of synonyms given by another method.

2.3   Discovery of Similar Words in a Dictionary

2.3.1   Introduction

We now propose a method for automatic synonym extraction in a monolingual dictionary [Sen01, BS01]. Our method uses a graph constructed from the dictionary and is based on the assumption that synonyms have many words in common in their definitions and are used in the definition of many common words. Our method is based on an algorithm that generalizes an algorithm initially proposed by Kleinberg for searching the Web [Kle99].

Starting from a dictionary, we first construct the associated *dictionary graph G*; each word of the dictionary is a vertex of the graph and there is an edge from u to v if v appears in the definition of u. Then, associated with a given query word w, we construct a *neighborhood graph G_w* which is the subgraph of G whose vertices are those pointed to by w or pointing to w. Finally, we look in the graph G_w for vertices that are similar to the vertex 2 in the structure graph

$$1 \longrightarrow 2 \longrightarrow 3$$

and choose these as synonyms. For this last step we use a similarity measure between vertices in graphs that was introduced in [BV02, Hey01].

The problem of searching synonyms is similar to that of searching similar pages on the Web; a problem that is dealt with in [Kle99] and [DH99]. In these references, similar pages are found by searching authoritative pages in a subgraph focused on the original page. Authoritative pages are pages that are similar to the vertex "authority" in the structure graph

$$\text{hub} \longrightarrow \text{authority.}$$

We ran the same method on the dictionary graph and obtained lists of good hubs and good authorities of the neighborhood graph. There were duplicates in these lists but not all good synonyms were duplicated. Neither authorities nor hubs appear to be the right concepts for discovering synonyms.

In the next section, we describe our method in some detail. In Section 2.3.3, we briefly survey two other methods that are used for comparison. We then describe in Section 2.3.4 how we have constructed a dictionary graph from *Webster's dictionary*. In the last section we compare all methods on the following words chosen for their variety: **disappear**, **parallelogram**, **sugar**, and **science**.

2.3.2   A Generalization of Kleinberg's Method

In [Kle99], Jon Kleinberg proposes a method for identifying Web pages that are good *hubs* or good *authorities* for a given query. For example, for the query "automobile makers", the home pages of Ford, Toyota, and other car makers are good authorities, whereas Web pages that list these homepages are good hubs. In order to identify hubs and authorities, Kleinberg's method exploits the natural graph structure of the Web in which each Web page is a vertex and there is an edge from vertex a to vertex b if page a points to page b. Associated with any given query word w, the method first constructs a "focused subgraph" G_w analogous to our neighborhood graph and then computes hub and authority scores for all vertices of G_w. These scores are obtained as the result of a converging iterative process. Initial hub and authority weights are all set to one, $x^1 = 1$ and $x^2 = 1$. These initial weights are then updated simultaneously according to a mutually reinforcing rule: the hub scores of the vertex i, x_i^1, is set equal to the sum of the authority scores of all vertices pointed to by i and, similarly, the authority scores of the vertex j, x_j^2, is set equal to the sum of the hub scores of all vertices pointing to j. Let M_w be the adjacency matrix associated with G_w. The updating equations can be written as

$$\begin{pmatrix} x^1 \\ x^2 \end{pmatrix}_{t+1} = \begin{pmatrix} 0 & M_w \\ M_w^T & 0 \end{pmatrix} \begin{pmatrix} x^1 \\ x^2 \end{pmatrix}_t \qquad t = 0, 1, \dots,$$

It can be shown that under weak conditions the normalized vector x^1 (respectively, x^2) converges to the normalized principal eigenvector of $M_w M_w^T$ (respectively, $M_w^T M_w$).

The authority score of a vertex v in a graph G can be seen as a similarity measure between v in G and vertex 2 in the graph

$$1 \longrightarrow 2.$$

Similarly, the hub score of v can be seen as a measure of similarity between v in G and vertex 1 in the same structure graph. As presented in [BV02, Hey01], this measure of similarity can be generalized to graphs that are different from the authority-hub structure graph. We describe below an extension of the method to a structure graph with three vertices and illustrate an application of this extension to synonym extraction.

Let G be a dictionary graph. The neighborhood graph of a word w is constructed with the words that appear in the definition of w and those that use w in their definition. Because of this, the word w in G_w is similar to the vertex 2 in the structure graph (denoted P_3)

$$1 \longrightarrow 2 \longrightarrow 3.$$

For instance, Figure 2.4 shows a part of the neighborhood graph of **likely**. The words **probable** and **likely** in the neighborhood graph are similar to the vertex 2 in P_3. The words **truthy** and **belief** are similar to, respectively, vertices 1 and 3. We say that a vertex is similar to vertex 2 of the preceding graph if it points to vertices that are similar to vertex 3 and if it is pointed to by vertices that are similar to vertex 1. This mutually reinforcing definition is analogous to Kleinberg's definitions of hubs and authorities.

The similarity between vertices in graphs can be computed as follows. With every vertex i of G_w we associate three scores (as many scores as there are vertices in the structure graph) x_i^1, x_i^2, and x_i^3 and initially set them equal to one. We then iteratively update the scores according to the following mutually reinforcing rule. The scores x_i^1 are set equal to the sum of the scores x_j^2 of all vertices j pointed to by

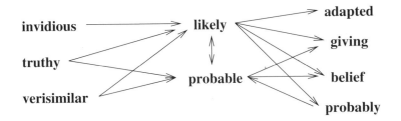

Figure 2.4. Subgraph of the neighborhood graph of **likely**.

i; the scores x_i^2 are set equal to the sum of the scores x_j^3 of vertices pointed to by i and the scores x_j^1 of vertices pointing to i; finally, the scores x_i^3 are set equal to the sum of the scores x_j^2 of vertices pointing to i. At each step, the scores are updated simultaneously and are subsequently normalized; $x^k \leftarrow x^k / \|x^k\|$ ($k = 1, 2, 3$). It can be shown that when this process converges, the normalized vector score x^2 converges to the normalized principal eigenvector of the matrix $M_w M_w^T + M_w^T M_w$. Thus our list of synonyms can be obtained by ranking in decreasing order the entries of the principal eigenvalue of $M_w M_w^T + M_w^T M_w$.

2.3.3   Other Methods

In this section, we briefly describe two synonym extraction methods that are compared to our method on a selection of four words.

The Distance Method

One possible way of defining a synonym distance is to declare that two words are close to being synonyms if they appear in the definition of many common words and have many common words in their definition. A way of formalizing this is to define a distance between two words by counting the number of words that appear in one of the definitions but not in both, and add to this the number of words that use one of the words but not both in their definition. Let A be the adjacency matrix of the dictionary graph, and i and j be the vertices associated with two words. The distance between i and j can be expressed as

$$d(i, j) = \|(A_{i,\cdot} - A_{j,\cdot})\|_1 + \|(A_{\cdot,i} - A_{\cdot,j})^T\|_1$$

where $\| \cdot \|_1$ is the l_1 vector norm. For a given word i we may compute $d(i, j)$ for all j and sort the words according to increasing distance.

Unlike the other methods presented in this chapter, we can apply this algorithm directly to the entire dictionary graph rather than to the neighborhood graph. This does, however, give very bad results: the first two synonyms of **sugar** in the dictionary graph constructed from *Webster's Dictionary* are **pigwidgeon** and **ivoride**. We show in Section 2.3.5 that much better results are achieved if we use the neighborhood graph.

ArcRank

ArcRank is a method introduced by Jan Jannink and Gio Wiederhold for building a thesaurus [JW99]; their intent was not to find synonyms but related words. An online version of their algorithm can be run from http://skeptic.stanford.edu/data/ (this online version also uses the 1913 *Webster's Dictionary* and the comparison with our results is therefore meaningful).

The method is based on the PageRank algorithm, used by the Web search engine Google and described in [BP98]. PageRank assigns a ranking to each vertex of the dictionary graph in the following way. All vertices start with identical initial ranking and then iteratively distribute it to the vertices they point to, while receiving the sum of the ranks from vertices that are pointed to them. Under conditions that are often satisfied in practice, the normalized ranking converges to a stationary distribution corresponding to the principal eigenvector of the adjacency matrix of the graph. This algorithm is actually slightly modified so that sources (nodes with no incoming edges, i.e., words not used in any definition) and sinks (nodes with no outgoing edges, i.e., words not defined) are not assigned extreme rankings.

ArcRank assigns a ranking to each edge according to the ranking of its vertices. If $|a_s|$ is the number of outgoing edges from vertex s and p_t is the page rank of vertex t, then the edge relevance of (s, t) is defined by

$$r_{s,t} = \frac{p_s/|a_s|}{p_t} .$$

Edge relevances are then converted into rankings. Those rankings are computed only once. When looking for words related to some word w, first select the edges starting from or arriving at w which have the best rankings and extract the corresponding incident vertices.

2.3.4   Dictionary Graph

Before proceeding to the description of our experiments, we describe how we constructed the dictionary graph. We used the Online Plain Text English Dictionary [OPT00] which is based on the "Project Gutenberg Etext of Webster's Unabridged Dictionary" which is in turn based on the 1913 US *Webster's Unabridged Dictionary*. The dictionary consists of 27 HTML files (one for each letter of the alphabet, and one for several additions). These files are available from the web site http://www.gutenberg.net/. In order to obtain the dictionary graph several choices had to be made.

- Some words defined in *Webster's Dictionary* are multiwords (e.g., **All Saints**, **Surinam toad**). We did not include these words in the graph since there is no simple way to decide, when the words are found side by side, whether they should be interpreted as single words or as a multiword (for instance, **at one** is defined but the two words **at** and **one** appear several times side by side in the dictionary in their usual meanings).

- Some head words of definitions were prefixes or suffixes (e.g., **un-**, **-ous**); these were excluded from the graph.

- Many words have several meanings and are head words of multiple definitions. Because, once more, it is not possible to determine which meaning of a word is employed in a definition, we gathered the definitions of a word into a single one.

- The recognition of derived forms of a word in a definition is also a problem. We dealt with the cases of regular and semiregular plurals (e.g., **daisies**, **albatrosses**) and regular verbs, assuming that irregular forms of nouns or verbs (e.g., **oxen**, **sought**) had entries in the dictionary.

- All accentuated characters were replaced in the HTML file by a \ (e.g., **proven\al**, **cr\che**). We included these words, keeping the \.

- There are many misspelled words in the dictionary, since it has been built by scanning the paper edition and processing it with an OCR software. We did not take these mistakes into account.

Because of the above remarks, the graph is far from being a precise graph of semantic relationships. For example, 13,396 lexical units are used in the definitions but are not defined. These include numbers (e.g., **14159265**, **14th**) and mathematical and chemical symbols (e.g., **x3**, **fe3o4**). When this kind of lexemes, which are not real words, is excluded, 12,461 words remain: proper names (e.g., **California**, **Aaron**), misspelled words (e.g., **aligator**, **abudance**), existing but undefined words (e.g., **snakelike**, **unwound**), or abbreviations (e.g., **adj**, **etc**).

The resulting graph has 112,169 vertices and 1,398,424 edges. It can be downloaded from `http://www.eleves.ens.fr:8080/home/senellar/st age_maitrise/graphe`. We analyzed several features of the graph: connectivity and strong connectivity, number of connected components, distribution of connected components, degree distributions, graph diameter, and so on. Our findings are reported in [Sen01].

We also decided to exclude too-frequent words in the construction of neighborhood graphs, that is, words that appear in more than L definitions (best results were obtained for $L \approx 1,000$). (The most commonly occurring words and their number of occurrences are: **of**: 68,187, **a**: 47,500, **the**: 43,760, **or**: 41,496, **to**: 31,957, **in**: 23,999, **as**: 22,529, **and**: 16,781, **an**: 14,027, **by**: 12,468, **one**: 12,216, **with**: 10,944, **which**: 10,446, **is**: 8,488, **for**: 8,188, **see**: 8,067, **from**: 7,964, **being**: 6,683, **who**: 6,163, **that**: 6,090).

2.3.5 Results

In order to be able to compare the different methods and to evaluate their relevance, we examine the first 10 results given by each of them for four words, chosen for their variety:

1. **disappear**: a word with various synonyms such as **vanish**;

2. **parallelogram**: a very specific word with no true synonyms but with some similar words: **quadrilateral**, **square**, **rectangle**, **rhomb**, …;

3. **sugar**: a common word with different meanings (in chemistry, cooking, dietetics, …). One can expect **glucose** as a candidate; and

4. **science**: a common and vague word. It is hard to say what to expect as a synonym. Perhaps **knowledge** is the best option.

Words in the English language belong to different parts of speech: nouns, verbs, adjectives, adverbs, prepositions, and so on. It is natural, when looking for a synonym of a word, to get only words of the same type. *Websters's Dictionary* provides the part of speech for each word. But this presentation has not been standardized and we counted not less than 305 different categories. We have chosen to select 5 types: nouns, adjectives, adverbs, verbs, others (including articles, conjunctions, and interjections) and have transformed the 305 categories into combinations of these types. A word may, of course, belong to different types. Thus, when looking for synonyms, we have excluded from the list all words that do not have a common part of speech with our word. This technique may be applied with all synonym extraction methods but since we did not implement ArcRank, we did not use it for ArcRank. In fact, the gain is not huge, because many words in English have several grammatical natures. For instance, **adagio** or **tete-a-tete** are at the same time nouns, adjectives, and adverbs.

We have also included lists of synonyms coming from WordNet [Wor98], which is handmade. The order of appearance of the words for this last source is arbitrary, whereas it is well defined for the distance method and for our method. The results given by the Web interface implementing ArcRank are two rankings, one for words pointed to by and one for words pointed to. We have interleaved them into one ranking. We have not kept the query word in the list of synonyms, in as much as this is only useful for our method, where it is interesting to note that in every example with which we have experimented, the original word appeared as the first word of the list (a point that tends to give credit to the method).

In order to have an objective evaluation of the different methods, we asked a sample of 21 persons to give a mark (from 0 to 10, 10 being the best one) to the lists of synonyms, according to their relevance to synonymy. The lists were, of course, presented in random order for each word. Figures 2.5 through 2.8 give the results.

Concerning **disappear**, the distance method and our method do pretty well: **vanish**, **cease**, **fade**, **die**, **pass**, **dissipate**, **faint** are very relevant (one must not forget that verbs necessarily appear without their postposition); **dissipate** or **faint** are relevant too. However, some words such as **light** or **port** are completely irrelevant, but they appear only in the sixth, seventh, or eigth position. If we compare these two methods, we observe that our method is better: an important synonym such as **pass** has a good ranking, whereas **port** or **appear** fall from the top 10 words. It is hard to explain this phenomenon, but we can say that the mutually reinforcing aspect of our method is apparently a positive point. On the contrary, ArcRank

	Distance	Our method	ArcRank	Wordnet
1	vanish	vanish	epidemic	vanish
2	wear	pass	disappearing	go away
3	die	die	port	end
4	sail	wear	dissipate	finish
5	faint	faint	cease	terminate
6	light	fade	eat	cease
7	port	sail	gradually	
8	absorb	light	instrumental	
9	appear	dissipate	darkness	
10	cease	cease	efface	
Mark	3.6	6.3	1.2	7.5
Std dev.	1.8	1.7	1.2	1.4

Figure 2.5. Proposed synonyms for **disappear**.

	Distance	Our method	ArcRank	Wordnet
1	square	square	quadrilateral	quadrilateral
2	parallel	rhomb	gnomon	quadrangle
3	rhomb	parallel	right-lined	tetragon
4	prism	figure	rectangle	
5	figure	prism	consequently	
6	equal	equal	parallelepiped	
7	quadrilateral	opposite	parallel	
8	opposite	angles	cylinder	
9	altitude	quadrilateral	popular	
10	parallelepiped	rectangle	prism	
Mark	4.6	4.8	3.3	6.3
Std dev.	2.7	2.5	2.2	2.5

Figure 2.6. Proposed synonyms for **parallelogram**.

gives rather poor results with words such as **eat**, **instrumental**, or **epidemic** that are imprecise.

Because the neighborhood graph of **parallelogram** is rather small (30 vertices), the first two algorithms give similar results, which are not absurd: **square**, **rhomb**, **quadrilateral**, **rectangle**, **figure** are rather interesting. Other words are less relevant but still are in the semantic domain of **parallelogram**. ArcRank, which also works on the same subgraph, does not give as interesting words, although **gnomon** makes its appearance, since **consequently** or **popular** are irrelevant. It is interesting to note that Wordnet is less rich here because it focuses on a particular aspect (**quadrilateral**).

Once more, the results given by ArcRank for **sugar** are mainly irrelevant (**property**, **grocer**). Our method is again better than the distance method: **starch**, **sucrose**, **sweet**, **dextrose**, **glucose**, and **lactose** are highly relevant words, even if the first given near-synonym (**cane**) is not as good. Its given mark is even better than for Wordnet.

	Distance	Our method	ArcRank	Wordnet
1	juice	cane	granulation	sweetening
2	starch	starch	shrub	sweetener
2	cane	sucrose	sucrose	carbohydrate
4	milk	milk	preserve	saccharide
5	molasses	sweet	honeyed	organic compound
6	sucrose	dextrose	property	saccarify
7	wax	molasses	sorghum	sweeten
8	root	juice	grocer	dulcify
9	crystalline	glucose	acetate	edulcorate
10	confection	lactose	saccharine	dulcorate
Mark	3.9	6.3	4.3	6.2
Std dev.	2.0	2.4	2.3	2.9

Figure 2.7. Proposed synonyms for **sugar**.

	Distance	Our method	ArcRank	Wordnet
1	art	art	formulate	knowledge domain
2	branch	branch	arithmetic	knowledge base
3	nature	law	systematize	discipline
4	law	study	scientific	subject
5	knowledge	practice	knowledge	subject area
6	principle	natural	geometry	subject field
7	life	knowledge	philosophical	field
8	natural	learning	learning	field of study
9	electricity	theory	expertness	ability
10	biology	principle	mathematics	power
Mark	3.6	4.4	3.2	7.1
Std dev.	2.0	2.5	2.9	2.6

Figure 2.8. Proposed synonyms for **science**.

The results for **science** are perhaps the most difficult to analyze. The distance method and ours are comparable. ArcRank gives perhaps better results than for other words but is still poorer than the two other methods.

To conclude, the first two algorithms give interesting and relevant words, whereas it is clear that ArcRank is not adapted to the search for synonyms. The variation of Kleinberg's algorithm and its mutually reinforcing relationship demonstrate its superiority on the basic distance method, even if the difference is not obvious for all words. The quality of the results obtained with these different methods is still quite different from that of handmade dictionaries such as Wordnet. Still, these automatic techniques show their interest, since they present more complete aspects of a word than handmade dictionaries. They can profitably be used to broaden a topic (see the example of **parallelogram**) and to help with the compilation of synonym dictionaries.

2.3.6   Future Perspectives

A first immediate improvement of our method would be to work on a larger subgraph than the neighborhood subgraph. The neighborhood graph we have introduced may be rather small, and may therefore not include important near-synonyms. A good example is **ox** of which **cow** seems to be a good synonym. Unfortunately, **ox** does not appear in the definition of **cow**, neither does the latter appear in the definition of the former. Thus the methods described above cannot find this word. Larger neighborhood graphs could be obtained either as Kleinberg does in [Kle99] for searching similar pages on the Web, or as Dean and Henziger do in [DH99] for the same purpose. However, such subgraphs are no longer focused on the original word. That implies that our variation of Kleinberg's algorithm "forgets" the original word and may produce irrelevant results. When we use the vicinity graph of Dean and Henziger, we obtain a few interesting results with specific words: for example, **trapezoid** appears as a near-synonym of **parallelogram** or **cow** as a near-synonym of **ox**. Yet there are also many degradations of performance for more general words. Perhaps a choice of neighborhood graph that depends on the word itself would be appropriate. For instance, the extended vicinity graph may be used for words whose neighborhood graph has less than a fixed number of vertices, or for words whose incoming degree is small, or for words that do not belong to the largest connected component of the dictionary graph.

One may wonder whether the results obtained are specific to *Webster's Dictionary* or whether the same methods could work on other dictionaries (using domain-specific dictionaries could, for instance, generate domain-specific thesauri, the value of which was mentioned in Section 2.2), in English or in other languages. Although the latter is most likely since our techniques were not designed for the particular graph we worked on, there will undoubtedly be differences with other languages. For example, in French, postpositions do not exist and thus verbs have fewer different meanings than in English. Besides, it is much rarer in French to have the same word for a noun and a verb than it is in English. Furthermore, the way words are defined varies from language to language. This seems to be an interesting research direction.

2.4   Conclusion

A number of different methods exist for the automatic discovery of similar words. Most of these methods are based on various text corpora and three of these are described in this chapter. Each of them may be more or less adapted to a specific problem (for instance, Crouch's techniques are more adapted to infrequent words than SEXTANT). We have also described the use of a more structured source - a monolingual dictionary - for the discovery of similar words. None of these methods is perfect and in fact none of them favorably competes with handmade dictionaries in terms of liability. Computer-written thesauri have, however, other advantages such as their ease of being built and rebuilt. The integration of different methods, with their own pros and cons, should be an interesting research direction to look at for designing successful methods. For it is most unlikely that a single straightforward technique may solve the issue of the discovery of similar words.

Another problem of the methods presented is the vagueness of the notion of "similar word" they use. Depending on the context, this notion may or may not include the notion of synonyms, near-synonyms, antonyms, hyponyms, and so on. The distinction between these very different notions by automatic means is a challenging problem that should be addressed to make it possible to build thesauri in a completely automatic way.

References

[BP98]   S. Brin and L. Page.The anatomy of a large-scale hypertextual Web search engine.*Computer Networks and ISDN Systems*, 30(1–7):107–117, 1998.

[BS01]   V.D. Blondel and P.P. Senellart.Automatic extraction of synonyms in a dictionary.Technical Report 89, Université catholique de Louvain, Louvain-la-neuve, Belgium, 2001.Presented at the Text Mining Workshop 2002 in Arlington, VA.

[BV02]   V.D. Blondel and P. Van Dooren.A measure of graph similarity between graph vertices.Technical Report, Université catholique de Louvain, Louvain-la-neuve, Belgium, 2002.

[CL92]   H. Chen and K.J. Lynch.Automatic construction of networks of concepts characterizing document databases.*IEEE Transactions on Systems, Man and Cybernetics*, 22(5):885–902, 1992.

[Cro90]   C.J. Crouch.An approach to the automatic construction of global thesauri.*Information Processing and Management*, 26:629–640, 1990.

[DH99]   J. Dean and M.R. Henzinger.Finding related pages in the World Wide Web.*WWW8 / Computer Networks*, 31(11-16):1467–1479, 1999.

[Gre93]   G. Grefenstette.Automatic thesaurus generation from raw text using knowledge-poor techniques.In *Making Sense of Words. Ninth Annual Conference of the UW Centre for the New OED and Text Research. 9*, 1993.

[Gre94]   G. Grefenstette.*Explorations in Automatic Thesaurus Discovery*.Kluwer Academic, Boston, 1994.

[Hey01]   M. Heymans.Extraction d'information dans les graphes, et application aux moteurs de recherche sur internet, Jun 2001.Université Catholique de Louvain, Faculté des Sciences Appliquées, Département d'Ingénierie Mathématique.

[JW99]   J. Jannink and G. Wiederhold.Thesaurus entry extraction from an on-line dictionary.In *Proceedings of Fusion '99*, Sunnyvale, CA, Jul 1999.

[Kle99]   J.M. Kleinberg.Authoritative sources in a hyperlinked environment.*Journal of the ACM*, 46(5):604–632, 1999.

[OPT00]   The online plain text english dictionary, 2000.http://msowww.anu.edu.au/~ralph/OPTED/.

[Sen01]   P. P. Senellart.Extraction of information in large graphs. Automatic search for synonyms.Technical Report 90, Université catholique de Louvain, Louvain-la-neuve, Belgium, 2001.

[SYY75]   G. Salton, C.S. Yang, and C.T. Yu.A theory of term importance in automatic text analysis.*Journal of the American Society for Information Science*, 26(1):33–44, 1975.

[Tur01]   P. D. Turney.Mining the Web for synonyms: PMI-IR versus LSA on TOEFL.In *Proceedings of the European Conference on Machine Learning*, pages 491–502, 2001.

[Wor98]   Wordnet 1.6, 1998.`http://www.cogsci.princeton.edu/~wn/`.

3

Simultaneous Clustering and Dynamic Keyword Weighting for Text Documents

Hichem Frigui
Olfa Nasraoui

Overview

In this chapter, we propose a new approach to unsupervised text document categorization based on a coupled process of clustering and cluster-dependent keyword weighting. The proposed algorithm is based on the K-Means clustering algorithm. Hence it is computationally and implementationally simple. Moreover, it learns a different set of keyword weights for each cluster. This means that, as a by-product of the clustering process, each document cluster will be characterized by a possibly different set of keywords. The cluster-dependent keyword weights have two advantages: they help in partitioning the document collection into more meaningful categories; and they can be used to automatically generate a compact description of each cluster in terms of not only the attribute *values*, but also their *relevance*. In particular, for the case of *text* data, this approach can be used to automatically annotate the documents. We also extend the proposed approach to handle the inherent fuzziness in text documents, by automatically generating fuzzy or soft labels instead of hard all-or-nothing categorization. This means that a text document can belong to *several* categories with different degrees. The proposed approach can handle noise documents elegantly by automatically designating one or two *noise magnet* clusters that grab most outliers away from the other clusters. The performance of the proposed algorithm is illustrated by using it to cluster real text document collections.

3.1 Introduction

Clustering is an important task that is performed as part of many text mining and information retrieval systems. Clustering can be used for efficiently finding the nearest neighbors of a document [BL85], for improving the precision or recall in information retrieval systems [vR79, Kow97], for aid in browsing a collection of documents [CKPT92], for the organization of search engine results [ZEMK97], and lately for the personalization of search engine results [Mla99].

Most current document clustering approaches work with what is known as the vector-space model, where each document is represented by a vector in the term-space. The latter generally consists of the keywords important to the document collection. For instance, the respective Term Frequencies (TF) [Kor77] in a given document can be used to form a vector model for this document. In order to discount frequent words with little discriminating power, each term/word can be weighted based on its Inverse Document Frequency (IDF) [Kor77, Mla99] in the document collection. However, the distribution of words in most real document collections can vary drastically from one group of documents to another. Hence relying solely on the IDF for keyword selection can be inappropriate and can severely degrade the results of clustering and/or any other learning tasks that follow it. For instance, a group of "News" documents and a group of "Business" documents are expected to have different sets of important keywords. Now, if the documents have already been manually preclassified into distinct categories, it would be trivial to select a different set of keywords for each category based on IDF. However, for large dynamic document collections, such as the case of World Wide Web documents, this manual classification is impractical, hence the need for automatic or unsupervised classification/clustering that can handle categories that differ widely in their best keyword sets. Unfortunately, it is not possible to differentiate between different sets of keywords, unless the documents have already been categorized. This means that in an unsupervised mode, both the categories and their respective keyword sets need to be discovered *simultaneously*. Selecting and weighting subsets of keywords in text documents is similar to the problem of feature selection and weighting in pattern recognition and data mining. The problem of selecting the best subset of features or attributes constitutes an important part of the design of good learning algorithms for real-world tasks. Irrelevant features can significantly degrade the generalization performance of these algorithms. In fact, even if the data samples have already been classified into known classes, it is generally preferable to model each complex class by several simple subclasses or clusters, and to use a different set of feature weights for each cluster. This can help in classifying new documents into one of the preexisting categories. So far, the problem of clustering and feature selection have been treated rather independently or in a wrapper kind of approach [AD91, KR92, RK92, JKP94, Ska94, KS95], but rarely coupled together to achieve the same objective.

In [FN00] we have presented a new algorithm, called Simultaneous Clustering and Attribute Discrimination (SCAD), that performs clustering and feature weighting *simultaneously*. When used as part of a supervised or unsupervised learning system, SCAD offers several advantages. First, its *continuous* feature weighting provides a much richer feature relevance representation than binary feature selection. Second, SCAD learns a *different* feature relevance representation for each cluster in an *unsupervised* manner. However, SCAD was intended for use with data lying in some Euclidean space, and the distance measure used was the Euclidean distance. For the special case of text documents, it is well known that the Euclidean distance is not appropriate, and other measures such as the cosine similarity or Jaccard index are better suited to assess the similarity/dissimilarity between documents.

In this chapter, we extend SCAD to *simultaneous text* document clustering and *dynamic category-dependent* keyword set weighting. This new approach to text clustering, that we call "Simultaneous KeyWord Identification and Clustering of text documents" or *SKWIC*, is both conceptually and computationally simple, and offers the following advantages compared to existing document clustering techniques. First, its *continuous* term weighting provides a much richer feature relevance representation than binary feature selection: Not all terms are considered *equally* relevant in a *single* category of text documents. This is especially true when the number of keywords is large. For example, one would expect the word "playoff" to be more important than the word "program" to distinguish a group of "sports" documents. Second, a given term is not considered *equally* relevant in *all* categories: For instance, the word "film" may be more relevant to a group of "entertainment" related documents than to a group of "sports" documents. Finally, SKWIC *learns* a *different* set of term weights for each cluster in an *unsupervised* manner.

We also extend the proposed approach to handle the inherent fuzziness in text documents, by automatically generating fuzzy or soft labels instead of single-label categorization. This means that a text document can belong to *several* categories with different degrees.

By virtue of the dynamic keyword weighting, and its continuous interaction with distance and membership computations, the proposed approach is able to handle noise documents elegantly by automatically designating one or two *noise magnet* clusters that grab most outliers away from the other clusters.

The organization of the rest of the chapter is as follows. In Section 3.2, we present the criterion for *SKWIC*, and derive necessary conditions to update the term weights. In Section 3.3, we present an alternative clustering technique, *Fuzzy SKWIC*, that provides richer *soft* document partitions. In Section 3.4, we explain how our approach achieves *robustness* to outliers in the data set. In Section 3.5, we illustrate the performance of SKWIC in unsupervised categorization of several text collections. Finally, Section 3.6 contains the summary conclusions.

3.2 Simultaneous Clustering and Term Weighting of Text Documents

SCAD [FN00] was formulated based on Euclidean distance. However, for many data mining applications such as clustering *text* documents and other *high-dimensional* data sets, the Euclidean distance measure is not appropriate. In general, the Euclidean distance is not a good measure for document categorization. This is due mainly to the high dimensionality of the problem, and the fact that two documents may not be considered similar if keywords are missing in both documents. More appropriate for this application is the cosine similarity measure [Kor77] between document frequency vectors $\mathbf{x_i}$ and $\mathbf{x_j}$ defined on a vocabulary of n terms,

$$S(x_i, x_j) = \frac{\sum_{k=1}^{n} x_{ik} \times x_{jk}}{\sqrt{\sum_{k=1}^{n} x_{ik}^2} \sqrt{\sum_{k=1}^{n} x_{jk}^2}}. \tag{3.1}$$

In order to be able to extend SCAD's criterion function for the case when another dissimilarity measure is employed, we only require the ability to decompose the dissimilarity measure across the different attribute directions. In this work, we attempt to decouple a dissimilarity based on the cosine similarity measure. We accomplish this by defining the dissimilarity between document $\mathbf{x}_j$ and the ith cluster center vector as follows.

$$\tilde{D}_{wc_{ij}} = \sum_{k=1}^{n} v_{ik} D_{wc_{ij}}^k, \tag{3.2}$$

which is the weighted aggregate sum of cosine-based distances along the individual dimensions, where

$$D_{wc_{ij}}^k = \frac{1}{n} - (x_{jk}.c_{ik}), \tag{3.3}$$

n is the total number of terms in a collection of N documents, c_{ik} is the kth component of the ith cluster center vector, and $\mathbf{V} = [v_{ik}]$ is the relevance weight of keyword k in cluster i. Note that the individual products are not normalized in Eq. (3.2) because it is assumed that the data vectors are normalized to unit length before they are clustered, and that all cluster centers are normalized after they are updated in each iteration.

SKWIC is designed to search for the optimal cluster centers $\mathbf{C}$ and the optimal set of feature weights $\mathbf{V}$ simultaneously. Each cluster i is allowed to have its own set of feature weights $\mathbf{V}_i = [v_{i1}, \ldots, v_{in}]$. We define the following objective function

$$\begin{aligned} J(\mathbf{C}, \mathbf{V}; \mathcal{X}) &= \sum_{i=1}^{C} \sum_{x_j \in \mathcal{X}_i} \sum_{k=1}^{n} v_{ik} D_{wc_{ij}}^k \\ &+ \sum_{i=1}^{C} \delta_i \sum_{k=1}^{n} v_{ik}^2, \end{aligned} \tag{3.4}$$

subject to

$$v_{ik} \in [0, 1] \; \forall \, i, \; k; \; \text{and} \; \sum_{k=1}^{n} v_{ik} = 1, \; \forall \, i. \tag{3.5}$$

The objective function in Eq. (3.4) has certain components. The first is the sum of distances or errors to the cluster centers. This component allows us to obtain compact clusters. It is minimized when only one keyword in each cluster is completely relevant, and all other keywords are irrelevant. The second component in Eq. (3.4) is the sum of the squared keyword weights. The global minimum of this component is achieved when all the keywords are equally weighted. When both components are combined and δ_i are chosen properly, the final partition will minimize the sum of intracluster weighted distances, where the keyword weights are optimized for each cluster.

To optimize J with respect to $\mathbf{V}$, we use the Lagrange multiplier technique, and obtain

$$J(\Lambda, \mathbf{V}) \;=\; \sum_{i=1}^{C} \sum_{x_j \in \mathcal{X}_i} \sum_{k=1}^{n} v_{ik} D_{wc_{ij}}^{k}$$

$$+ \sum_{i=1}^{C} \delta_i \sum_{k=1}^{n} v_{ik}^2 - \sum_{i=1}^{C} \lambda_i \left(\sum_{k=1}^{n} v_{ik} - 1 \right),$$

where $\Lambda = [\lambda_1, \dots, \lambda_c]^T$. Since the rows of $\mathbf{V}$ are independent of each other, we can reduce the above optimization problem to the following C independent problems

$$J_i(\lambda_i, \mathbf{V}_i) \;=\; \sum_{x_j \in \mathcal{X}_i} \sum_{k=1}^{n} v_{ik} D_{wc_{ij}}^{k}$$

$$+ \delta_i \sum_{k=1}^{n} v_{ik}^2 - \lambda_i \left(\sum_{k=1}^{n} v_{ik} - 1 \right),$$

$$\text{for } i = 1, \dots, C,$$

where $\mathbf{V}_i$ is the ith row of $\mathbf{V}$. By setting the gradient of J_i to zero, we obtain

$$\frac{\partial J_i(\lambda_i, \mathbf{V}_i)}{\partial \lambda_i} = \left(\sum_{k=1}^{n} v_{ik} - 1 \right) = 0, \tag{3.6}$$

and

$$\frac{\partial J_i(\lambda_i, \mathbf{V}_i)}{\partial v_{ik}} = \sum_{x_j \in \mathcal{X}_i} D_{wc_{ij}}^{k} + 2\delta_i v_{ik} - \lambda_i = 0. \tag{3.7}$$

Solving Eqs. (3.6) and (3.7) for v_{ik}, we obtain

$$v_{ik} = \frac{1}{n} + \frac{1}{2\delta_i} \sum_{x_j \in \mathcal{X}_i} \left[\frac{1}{n} \sum_{x_j \in \mathcal{X}_i} D^k_{wc_{ij}} - D^k_{wc_{ij}} \right]. \tag{3.8}$$

The first term in Eq. (3.8), $(1/n)$, is the default value if all attributes/keywords are treated equally, and no discrimination is performed. The second term is a bias that can be either positive or negative. It is positive for compact attributes where the distance along this dimension is, on the average, less than the total distance using all of the dimensions. If an attribute is very compact, compared to the other attributes, for most of the points that belong to a given cluster, then it is very relevant for that cluster. Note that it is possible for the individual termwise dissimilarities in Eq. (3.3) to become negative. This will simply emphasize that dimension further and will result in relatively larger attribute weights v_{ik} (see Eq. (3.8)). Moreover, the total aggregate dissimilarity in Eq. (3.2) can become negative. This also does not pose a problem because we partition the data based on minimum distance.

The choice of δ_i in Eq. (3.4) is important in the SKWIC algorithm since it reflects the importance of the second term relative to the first term. If δ_i is too small, then only one keyword in cluster i will be relevant and assigned a weight of one. All other words will be assigned zero weights. On the other hand, if δ_i is too large, then all words in cluster i will be relevant, and assigned equal weights of $1/n$. The values of δ_i should be chosen such that both terms are of the same order of magnitude. In all examples described in this chapter, we compute δ_i in iteration, t, using

$$\delta_i^{(t)} = K_\delta \frac{\sum\limits_{x_j \in \mathcal{X}_i} \sum\limits_{k=1}^{n} v_{ik}^{(t-1)} \left(D_{wc_{ij}}^{k^{(t-1)}} \right)}{\sum\limits_{k=1}^{n} \left(v_{ik}^{(t-1)} \right)^2}. \tag{3.9}$$

In Eq. (3.9), K_δ is a constant, and the superscript $(t-1)$ is used on u_{ij}, v_{ik}, and c_{ik} to denote their values in iteration $(t-1)$.

It should be noted that depending on the values of δ_i, the feature relevance values v_{ik} may not be confined to $[0,1]$. If this occurs very often, then it is an indication that the value of δ is too small, and that it should be increased (increase K_δ). On the other hand, if this occurs for a few clusters and only in a few iterations, then we adjust the negative feature relevance values as follows

$$v_{ik} \leftarrow v_{ik} + \left| \min_{k=1}^{n} v_{ik} \right| \text{ if } v_{ik} < 0. \tag{3.10}$$

It can also be shown that the cluster partition that minimizes J is the one that assigns each data sample to the cluster with *nearest* prototype/center, that is,

$$\mathcal{X}_i = \left\{ \mathbf{x}_j | \tilde{D}_{wc_{ij}} \leq \tilde{D}_{wc_{kj}} \forall k \neq i \right\}, \tag{3.11}$$

where $\tilde{D}_{wc_{kj}}$ is the weighted aggregate cosine-based distance in Eq. (3.2), and ties are resolved arbitrarily.

It is not possible to minimize J with respect to the centers. Hence, we compute the new cluster centroids (as in the ordinary SCAD algorithm [FN00]) and normalize them to unit length to obtain the new cluster centers. We obtain two cases depending on the value of v_{ik}.

Case 1: $v_{ik} = 0$
In this case the kth feature is completely irrelevant relative to the ith cluster. Hence, regardless of the value of c_{ik}, the values of this feature will not contribute to the overall weighted distance computation. Therefore, in this situation, any arbitrary value can be chosen for c_{ik}. In practice, we set $c_{ik} = 0$.

Case 2: $v_{ik} \neq 0$
For the case when the kth feature has some relevance to the ith cluster, the center reduces to

$$c_{ik} = \frac{\sum\limits_{x_j \in \mathcal{X}_i} x_{jk}}{\sum\limits_{x_j \in \mathcal{X}_i} 1}.$$

To summarize, the update equation for the centers is

$$c_{ik} = \begin{cases} 0, & \text{if } v_{ik} = 0, \\ \dfrac{\sum\limits_{x_j \in \mathcal{X}_i} x_{jk}}{|\mathcal{X}_i|}, & \text{if } v_{ik} > 0. \end{cases} \tag{3.12}$$

Finally, we summarize the SKWIC algorithm for clustering a collection of N normalized document vectors defined over a vocabulary of n keywords.

Algorithm 3.1 Simultaneous Keyword Identification and Clustering of Text Documents (SKWIC)

Fix the number of clusters C;
Initialize the centers by randomly selecting C documents;
Initialize the partitions, $\mathcal{X}_i$, using (3.11) and equal feature weights ($1/n$);

REPEAT
   Compute $D^k_{wc_{ij}} = 1/n - (x_{jk}.c_{ik})$
    for $1 \leq i \leq C$, $1 \leq j \leq N$, and $1 \leq k \leq n$;
   Update the relevance weights v_{ik} by using (3.8);
   Compute $\tilde{D}_{wc_{ij}}$ for $1 \leq i \leq C$, $1 \leq j \leq N$, using (3.2);
   Update the cluster partition $\mathcal{X}_i$ by using (3.11);
   Update the centers by using (3.12);
   Update δ_i by using (3.9);
UNTIL (*centers stabilize*);

The feature weighting equations used in SKWIC may be likened to the estimation and use of a covariance matrix in an inner-product norm-induced metric [GK79] in various statistical pattern recognition techniques. However, the estimation of a covariance matrix does not really weight the attributes according to their relevance, and it relies on the assumption that the data have a multivariate Gaussian distribution. On the other hand, SKWIC is free of any such assumptions when estimating the feature weights. This means that SKWIC can be adapted to more general dissimilarity measures, such as was done in this chapter with the cosine-based dissimilarity.

3.3   Simultaneous Soft Clustering and Term Weighting of Text Documents

Documents in a collection can rarely be described as members of a single/exclusive category. In fact, most documents will tend to straddle the subject of two or more different subjects. Even manual classification is difficult and poor in this case, because each document is finally labeled into a single class, and this can drastically affect retrieval abilities once a classification model is built. Hard partitioning models such as K-Means and SKWIC are constrained to assign every document to a single cluster/category, and the final assignment is often poor in modeling documents that can be assigned to more than one category. Consequently they are expected to have limited capability for real large document collections. In this section, we present a technique to provide a *soft* unsupervised categorization of a collection of documents. By soft, it is meant that a given document must not be confined to a single category.

It is known that for complex data sets containing overlapping clusters, fuzzy/soft partitions model the data better than their crisp/hard counterparts. In particular, fuzzy memberships are richer than crisp memberships in describing the degrees of association of datapoints lying in the areas of overlap. Moreover, fuzzy partitions generally smooth the surface of the criterion function in the search space and, hence, make the optimization process less prone to local or suboptimal solutions. With a fuzzy partition, a datapoint $\mathbf{x}_j$ belongs to each cluster $\mathcal{X}_i$ to a varying degree called fuzzy membership u_{ij}. A fuzzy partition, usually represented by the $C \times N$ matrix $\mathbf{U} = [u_{ij}]$, is called a constrained fuzzy $C-$partition of $\mathcal{X}$ if the entries of $\mathbf{U}$ satisfy the following constraints [Bez81],

$$\begin{cases} u_{ij} \in [0, 1], & \forall i, \\ 0 < \displaystyle\sum_{j=1}^{N} u_{ij} < N, & \forall i, j, \\ \displaystyle\sum_{i=1}^{C} u_{ij} = 1, & \forall j. \end{cases} \quad (3.13)$$

Fuzzy SKWIC is designed to search for the optimal cluster centers $\mathbf{C}$, the optimal soft partitioning memberships $\mathbf{U}$, and the optimal set of feature weights $\mathbf{V}$, simultaneously. Each cluster i is allowed to have its own set of feature weights $\mathbf{V}_i = [v_{i1}, \ldots, v_{in}]$, and fuzzy membership degrees (u_{ij}) that define a fuzzy partition of the data set satisfying (3.13). We define the following objective function,

$$\begin{aligned} J(\mathbf{C}, \mathbf{U}, \mathbf{V}; \mathcal{X}) &= \sum_{i=1}^{C} \sum_{j=1}^{N} (u_{ij})^m \sum_{k=1}^{n} v_{ik} D_{wc_{ij}}^k \\ &+ \sum_{i=1}^{C} \delta_i \sum_{k=1}^{n} v_{ik}^2, \end{aligned} \quad (3.14)$$

subject to

$$v_{ik} \in [0, 1] \, \forall \, i, \, k; \text{ and } \sum_{k=1}^{n} v_{ik} = 1, \, \forall \, i. \quad (3.15)$$

The objective function in (3.14) has two components. One component is the sum of distances or errors to the cluster centers. This component allows us to obtain compact clusters. It is minimized when only one keyword in each cluster is completely relevant, and all other keywords are irrelevant. The othe component in Eq. (3.14) is the sum of the squared keyword weights. The global minimum of this component is achieved when all the keywords are equally weighted. When both components are combined and δ_i are chosen properly, the final partition will minimize the sum of intracluster weighted distances, where the keyword weights are optimized for each cluster.

To optimize J, with respect to $\mathbf{V}$, we use the Lagrange multiplier technique, and obtain

$$J(\Lambda, \mathbf{V}) = \sum_{i=1}^{C} \sum_{j=1}^{N} (u_{ij})^m \sum_{k=1}^{n} v_{ik} D_{wc_{ij}}^{k}$$
$$+ \sum_{i=1}^{C} \delta_i \sum_{k=1}^{n} v_{ik}^2 - \sum_{i=1}^{C} \lambda_i \left(\sum_{k=1}^{n} v_{ik} - 1 \right),$$

where $\Lambda = [\lambda_1, \ldots, \lambda_c]^t$. Since the rows of $\mathbf{V}$ are independent of each other, we can reduce the above optimization problem to the following C independent problems,

$$J_i(\lambda_i, \mathbf{V}_i) = \sum_{j=1}^{N} (u_{ij})^m \sum_{k=1}^{n} v_{ik} D_{wc_{ij}}^{k}$$
$$+ \delta_i \sum_{k=1}^{n} v_{ik}^2 - \lambda_i \left(\sum_{k=1}^{n} v_{ik} - 1 \right)$$
$$\text{for } i = 1, \ldots, C,$$

where $\mathbf{V}_i$ is the ith row of $\mathbf{V}$. By setting the gradient of J_i to zero, we obtain

$$\frac{\partial J_i(\lambda_i, \mathbf{V}_i)}{\partial \lambda_i} = \left(\sum_{k=1}^{n} v_{ik} - 1 \right) = 0, \tag{3.16}$$

and

$$\frac{\partial J_i(\lambda_i, \mathbf{V}_i)}{\partial v_{ik}} = \sum_{j=1}^{N} (u_{ij})^m D_{wc_{ij}}^{k} + 2\delta_i v_{ik} - \lambda_i = 0. \tag{3.17}$$

Solving (3.16) and (3.17) for v_{ik}, we obtain

$$v_{ik} = \frac{1}{n} + \frac{1}{2\delta_i} \sum_{j=1}^{N} (u_{ij})^m \left[\frac{\tilde{D}_{wc_{ij}}}{n} - D_{wc_{ij}}^{k} \right]. \tag{3.18}$$

The first term in (3.18), $(1/n)$, is the default value if all attributes/keywords are treated equally, and no discrimination is performed. The second term is a bias that can be either positive or negative. It is positive for compact attributes where the distance along this dimension is, on the average, less than the total distance using all of the dimensions. If an attribute is very compact, compared to the other attributes, for most of the points that belong to a given cluster (high u_{ij}), then it is very relevant for that cluster. Note that it is possible for the individual term-wise dissimilarities in (3.3) to become negative. This will simply emphasize that dimension further and will result in relatively larger attribute weights v_{ik} (see (3.18)).

The choice of δ_i in Eq. (3.14) is important in the Fuzzy SKWIC algorithm since it reflects the importance of the second term relative to the first term. If δ_i is too small, then only one keyword in cluster i will be relevant and assigned a weight of one. All other words will be assigned zero weights. On the other hand, if δ_i is too large, then all words in cluster i will be relevant, and assigned equal weights of $1/n$. The values of δ_i should be chosen such that both terms are of the same order of magnitude. In all examples described in this chapter, we compute δ_i in iteration, t, using

$$\delta_i^{(t)} = K_\delta \frac{\sum_{j=1}^{N} \left(u_{ij}^{(t-1)} \right)^m \sum_{k=1}^{n} v_{ik}^{(t-1)} \left(D_{wc_{ij}}^{k^{(t-1)}} \right)}{\sum_{k=1}^{n} \left(v_{ik}^{(t-1)} \right)^2}. \tag{3.19}$$

In (3.19), K_δ is a constant, and the superscript $(t-1)$ is used on u_{ij}, v_{ik}, and c_{ik} to denote their values in iteration $(t-1)$.

It should be noted that depending on the values of δ_i, the feature relevance values v_{ik} may not be confined to [0,1]. If this occurs very often, then it is an indication that the value of δ is too small, and that it should be increased (increase K_δ). However, if this occurs for few clusters and only in few iterations, then we adjust the negative feature relevance values as follows

$$v_{ik} \leftarrow v_{ik} + \left| \min_{k=1}^{n} v_{ik} \right| \text{ if } v_{ik} < 0. \tag{3.20}$$

Since the second term in (3.14) does not depend on u_{ij} explicitly, the update equation of the memberships is similar to that of the Fuzzy C Means, that is,

$$u_{ij} = \frac{1}{\sum_{k=1}^{C} \left(\frac{\tilde{D}_{wc_{ij}}}{\tilde{D}_{wc_{kj}}} \right)^{1/(m-1)}}. \tag{3.21}$$

The componentwise distance values $D_{wc_{ij}}^{k}$ in (3.3) can be negative, and hence the overall distance $D_{wc_{ij}}^{k}$ in (3.2) can become negative, which can affect the sign of the fuzzy memberships in (3.21). Therefore, we adjust the negative distance values as follows

$$\tilde{D}_{wc_{ij}} \leftarrow \tilde{D}_{wc_{ij}} + \left| \min_{i=1}^{C} \tilde{D}_{wc_{ij}} \right| \text{ if } \tilde{D}_{wc_{ij}} < 0. \tag{3.22}$$

Finally, the update equation for the centers which takes into account the soft memberships/partition is

$$c_{ik} = \begin{cases} 0, & \text{if } v_{ik} = 0, \\[2em] \dfrac{\displaystyle\sum_{j=1}^{N}(u_{ij})^m x_{jk}}{\displaystyle\sum_{j=1}^{N}(u_{ij})^m}, & \text{if } v_{ik} > 0. \end{cases} \qquad (3.23)$$

We summarize the Fuzzy SKWIC algorithm below.

Algorithm 3.2 Simultaneous Keyword Identification and Clustering of Text Documents (Fuzzy SKWIC)

Fix the number of clusters C;
Fix m, m $\in$ [1, ∞);
Initialize the centers by randomly selecting C documents;
Initialize the fuzzy partition matrix **U** *;*
REPEAT
  Compute $D^k_{wc_{ij}} = (1/n) - (x_{jk} \cdot c_{ik})$
    for $1 \leq i \leq C$, $1 \leq j \leq N$, and $1 \leq k \leq n$;
  Update the relevance weights v_{ik} by using (3.18);
  Adjust relevance weights v_{ik} by using (3.20);
  Compute $\tilde{D}_{wc_{ij}}$ for $1 \leq i \leq C$, $1 \leq j \leq N$, using (3.2);
  Adjust negative $\tilde{D}_{wc_{ij}}$ using (3.22);
  Update the partition matrix $\mathbf{U}^{(k)}$ by using (3.21);
  Update the centers by using (3.23);
  Update δ_i by using (3.19);
UNTIL (*centers stabilize*);

3.4   Robustness in the Presence of Noise Documents

When there are several documents that do not form a strong consensus or cluster (i.e., they are neither similar to each other nor to any of the other compact clusters), because our distances are confined in [0, 1], all outlier documents will have a maximal distance of 1. Hence, their effect on the objective functions in (3.4) and (3.14) is limited. This means that they cannot drastically influence the results for other clusters. This limited influence, by definition, makes our approach *robust* in the face of outliers and noise. In essence, this is similar to using a robust loss function, $\rho()$, in M-Estimators [Hub81, RL87].

Moreover, because the distance between outliers and all clusters is close to the maximal value of 1, if they happen to get assigned to any one of the clusters initialized with a seed that is close to the outliers, they will tend to pull all the keyword relevance weights to a low value in that cluster because of extreme averaging. This in turn will further bias the distance computations to this cluster to be small. As a result, this cluster will start acting like a magnet that continues to grab documents that are not very typical of any category towards it, and therefore keep growing. Only documents that are really similar to their cluster's centroid will remain in their own clusters, and hence avoid being pulled into the noise cluster. Consequently, designated *noise magnet* clusters will help in keeping the remaining clusters cleaner and their constituents more uniform.

We have observed the emergence of such *noise magnets* in every experiment that we performed.

3.5   Experimental Results

3.5.1   *Simulation Results on Four-Class Web Text Data*

Simulation Results with Hard Clustering

The first experiment illustrates the clustering results on a collection of text documents collected from the World Wide Web. Students were asked to collect 50 distinct documents from each of the following categories: news, business, entertainment, and sports. Thus the entire collection consisted of 200 documents. The documents' contents were preprocessed by eliminating stop words and stemming words to their root source. Then, the IDF [Kor77] of the terms were computed and sorted in descending order so that only the top 200 terms were chosen as final keywords. Finally each document was represented by the vector of its document frequencies, and this vector was normalized to unit length. Using $C = 4$ as the number of clusters, SKWIC converged after five iterations, resulting in a partition that closely resembles the distribution of the documents with respect to their true categories. The class distribution is shown in Table 3.1. Table 3.2 lists the six most relevant keywords for each cluster. As can be seen, the collection of terms receiving highest feature relevance weights in each cluster reflected the general topic of the category winning the majority of the documents that were assigned to the cluster. In addition, these cluster-dependent keywords can be used to provide a short summary for each cluster and to automatically annotate documents.

The partition of the Class 2 documents showed most of the error in assignment due to the mixed nature of some of the documents therein. For example, by looking at the excerpts (shown below) from the following documents from Class 2 (*entertainment*) that were assigned to Cluster 1 with relevant words relating to *business* as seen in Table 3.2, one can see that these documents are hard to classify into one category, and that the keywords present in the documents in this case have misled the clustering process.

Excerpt from Document 54:

The couple were together for 3-1/2 years before their highly publicized split last month. Now, their Ojai property is on the market for $2.75 million, the Los Angeles Times reported on Sunday. The pair bought the 10-acre Ojai property - complete with working avocado and citrus orchards - at the end of 1998. They also purchased a Hollywood Hills home for $1.7 million in June 1999, according to the Times.

Excerpt from Document 59:

The recommendation, approved last week by the joint strike committee for the Screen Actors Guild (SAG) and the American Federation of Television & Radio Artists (AFTRA), would have to be approved by the national boards of the unions to go into effect - a process that would take a month to complete. "Part of this is motivated by the awareness of actors who have been egregious about performing struck work and part of it is trying to recognize the 99.999% of members who have stuck together on this," SAG spokesman Greg Krizman said.

Excerpt from Document 78:

The Oxford-based quintet's acclaimed fourth release, "Kid A," opened at No. 1 with sales of 207,000 copies in the week ended Oct. 8, the group's Capitol Records label said Wednesday. The tally is more than four times the first-week sales of its previous album. The last Stateside No. 1 album from the U.K. was techno act Prodigy's "The Fat of the Land" in July 1997. That very same week, Radiohead's "OK Computer" opened at No. 21 with 51,000 units sold. It went on to sell 1.2 million copies in the United States.

The above excerpts further illustrate the inherent *fuzziness* in categorizing text documents, as the shown documents straddle the *business* and *entertainment* categories. In this case, it can be said that the baseline manual labeling was not accurate. Fuzzy or soft labels are desired for such documents, and these are illustrated in the next section.

	Cluster 1	Cluster 2	Cluster 3	Cluster 4
	(business)	(entertainment)	(news)	(sports)
class 1	45	2	3	0
class 2	9	31	4	6
class 3	1	1	47	1
class 4	0	0	4	46

Table 3.1. Distribution of the 50 Documents from Each Class into the Four Clusters Computed by SKWIC

Simulation Results with Soft Clustering

Using $C = 4$ as the number of clusters, and $m = 1.1$, Fuzzy SKWIC converged after 27 iterations, resulting in a partition that closely resembles the distribution of the documents with respect to their true categories.

Cluster # 1		Cluster # 2		Cluster # 3		Cluster # 4	
$v_{1(k)}$	$w_{(k)}$	$v_{2(k)}$	$w_{(k)}$	$v_{3(k)}$	$w_{(k)}$	$v_{4(k)}$	$w_{(k)}$
0.028	compani	0.031	film	0.009	polic	0.021	game
0.015	percent	0.012	star	0.008	nation	0.013	season
0.010	share	0.010	dai	0.008	state	0.012	open
0.010	expect	0.010	week	0.008	offici	0.009	york
0.009	market	0.009	peopl	0.008	sai	0.008	hit
0.008	stock	0.008	like	0.007	kill	0.008	run

Table 3.2. Term Relevance for the Top Six Relevant Words in Each Cluster computed by SKWIC

The class distribution is shown in Table 3.3 and the six most relevant keywords for each cluster are listed in Table 3.4. The highly relevant keywords (top two or three) are consistent with those obtained using the crisp version. The partition obtained using the fuzzy SKWIC (Table 3.3) is slightly better than the one obtained using the crisp SKWIC (Table 3.1). The partition of the Class 2 documents still shows the same number of classification errors as in the crisp case. However, a careful examination of the misclassified documents shows that these documents have high membership degrees in more than one cluster, and thus should not be assigned one simple label. Thus the class distribution in Table 3.3 would greatly improve if the groundtruth labeling were soft from the start. The following excerpt illustrates the soft labels that are automatically computed by Fuzzy SKWIC. They clearly show a document that is *Mostly about Entertainment, but Somewhat also relating to Business*. Hence in addition to relevant keywords that provide a *short summary* for each cluster, Fuzzy SKWIC can generate a richer soft labeling of the text documents that can aid in retrieval.

Excerpt from Document 70 [soft labels: Business = 85% ($u_{0j} = 0.853$), Entertainment= 14% ($u_{1j} = 0.140$), News = 0.5% ($u_{2j} = 0.005$), Sports = 0.3% ($u_{3j} = 0.003$)] :

LOS ANGELES (Reuters) – Ifilm and Pop.com, the would-be Web site backed by filmmakers Steven Spielberg, Ron Howard, and other Hollywood moguls, have ended talks to merge, according to an e-mail sent to Ifilm employees on Friday. "The companies will continue to enjoy many overlapping shareholder and personal relationships," the memo said. Industry observers said the founders of Pop.com, which has never aired a single show or launched its Web site, are looking for a graceful exit strategy out of the venture, which has been plagued by infighting and uncertainty about the company's direction and business plan.

3.5.2   Simulation Results on 20 Newsgroups Data

The second set of experiments is based on the 20 newsgroups data set [CMU]. This data set is a collection of 20,000 messages, collected from 20 different netnews newsgroups. One thousand messages from each of the 20 newsgroups were chosen at random and partitioned by newsgroup name. The list of newsgroups from which

	Cluster # 1	Cluster # 2	Cluster # 3	Cluster # 4
	(business)	(entertainment)	(news)	(sports)
class 1	48	1	1	0
class 2	7	31	5	7
class 3	2	1	47	0
class 4	0	0	3	47

Table 3.3. Distribution of the 50 Documents from Each Class into the Four Clusters Computed by Fuzzy SKWIC

Cluster # 1		Cluster # 2		Cluster # 3		Cluster # 4	
$v_{1(k)}$	$w_{(k)}$	$v_{2(k)}$	$w_{(k)}$	$v_{3(k)}$	$w_{(k)}$	$v_{4(k)}$	$w_{(k)}$
0.029	compani	0.031	film	0.016	polic	0.025	game
0.016	percent	0.012	star	0.011	govern	0.015	season
0.011	share	0.010	week	0.010	state	0.010	plai
0.010	expect	0.008	dai	0.009	offici	0.009	york
0.008	market	0.008	peopl	0.009	nation	0.009	open
0.008	stock	0.008	open	0.009	sai	0.009	run

Table 3.4. Term Relevance for the Top Six Relevant Words in Each Cluster Computed by Fuzzy SKWIC

the messages were chosen is shown in Table 3.5. The documents were first pre-processed: this included stripping each news message from the e-mail header and special tags, then eliminating stop-words and finally stemming words to their root form using the *rainbow* software package [Bow]. Next, words were sorted based on their IDF values. Finally, the number of keywords was reduced by selecting them based on setting a minimum threshold on their sorted IDF values, so as not to exceed a maximum number of words. Since several documents end up with none of the words that were selected, these documents are not considered for clustering. We first present a discussion of the results obtained on a subset of 2000 documents from the 20 newsgroups data set. This data set is called the *mini newsgroup data set* [Bow]. Then we discuss the results on the entire 20 newsgroups data set.

Simulation Results on Mini Newsgroups Data using SKWIC

After preprocessing, 449 words were selected based on IDF. Consequently, there were 1730 documents with at least one of these selected keywords. The documents were clustered by SKWIC into $C = 40$ clusters. Note that we arbitrarily chose this number because the actual messages may be categorized better with more clusters. In other words, there is no guarantee that the labeled documents really come from $K = 20$ different categories, since the labeling was done based on the newsgroup name. Moreover, there is no control over messages that may be sent to a particular newsgroup since their topic may differ from the majority in that newsgroup, or even be more similar to a completely different newsgroup.

Class	Class Descriptions	Class	Class Descriptions
1	alt.atheism	11	rec.sport.hockey
2	comp.graphics	12	sci.crypt
3	comp.os.ms-windows.misc	13	sci.electronics
4	comp.sys.ibm.pc.hardware	14	sci.med
5	comp.sys.mac.hardware	15	sci.space
6	comp.windows.x	16	soc.religion.christian
7	misc.forsale	17	talk.politics.guns
8	rec.autos	18	talk.politics.mideast
9	rec.motorcycles	19	talk.politics.misc
10	rec.sport.baseball	20	talk.religion.misc

Table 3.5. Twenty Class Descriptions

Table 3.6 shows the class distribution of the 40 clusters discovered by SKWIC. The columns correspond to the class indices that can be mapped to a complete class description using Table 3.5. In general, each row shows one or a few large values, which indicates that the algorithm succeeds in partitioning the majority of the same newsgroup documents into a few homogeneous clusters according to the specific nature of the documents.

Table 3.7 displays the cluster cardinalities, as well as the top 10 relevant keywords for each cluster, sorted in decreasing order of their relevance weights in each cluster. Note how the relevance weights may vary drastically between different clusters, and this has a significant effect on the weighted distance computations, and hence affects the final partitioning of the documents. By looking at which keywords have the highest relevance in a given cluster, and their relevance values, it is possible to roughly deduce the nature of the newsgroup messages that fall into one particular cluster. For example, some cluster keyword relevances seem to suggest a stream of discussions that are specific to either a certain event that occurred or to a particular issue that grabbed the attention of a subset of participants in a certain newsgroup. Consequently, it can also be seen how some of these clusters can be formed from documents from *distinct* newsgroups because the messages seemed to relate to similar issues that cross different newsgroups. Several such *mixed* clusters can be formed from documents that cross the boundary between different politics groups, between different religion groups, and even between both politics and religion groups, and so on.

Table 3.6 shows some clusters that include documents from different, yet *related*, newsgroups. For instance, Cluster No. 3 seems to group several documents (61) from all 5 comp. newsgroups (but with the majority from the *comp.graphics* newsgroup), as well as the sci.electronics (8) and sci.med (6), but surprisingly also some from soc.religion.christian (7) and some from talk.religion.misc (7). Table 3.7 lists the top 10 relevant keywords for this cluster that are indicative of the type of content in messages from the comp. and some of the sci. groups, but not necessarily the religion groups. For example, some of the sci.space documents assigned to this cluster speak about solar and lunar images, hence the affinity to graphics.

Another message from the talk.religion.misc newsgroup, was assigned to Cluster 3 because of some content relating to computers. It had the following quote in the sender's signature:*"A system admin's life is a sorry one. The only advantage he has over Emergency Room doctors is that malpractice suits are rare. On the other hand, ER doctors never have to deal with patients installing new versions of their own innards!"* Here is an excerpt from another message from the talk.religion.misc newsgroup, that was assigned to Cluster 3 because of the scientific rhetoric (which pulled it towards the comp. and sci. documents in Cluster 3):*"This, again, is a belief, not a scientific premise. The original thread referred specifically to 'scientific creationism. This means whatever theory or theories you propose must be able to be judged by the scientific method'."*

There were also several messages concentrating on a major event during that period (Waco's battle), that were assigned to Cluster No. 3, mainly because of the presence of one of the relevant keywords (*semi*). Here is one of the excerpts: *"in other words faith in a .357 is far stronger than faith in a God providing a miracle for his followers. Interesting. Now, if David Korresh was God, why couldn't he use lightning instead of semi-automatic rifles?"* This example illustrates a typical example where the same keyword (*semi*) may have different meanings depending on context.

Just as a cluster can group documents from several related newsgroups, a particular newsgroup may be split into two or more clusters according to the specific topic of the documents. For example, the rec.sport.hockey newsgroup is split over Clusters No. 20 and 21, as can be seen in Table 3.6. Cluster 20 contains more documents from the rec.sport.baseball group, while Cluster No. 21 is more specific to hockey. Table 3.7 reveals completely different keyword distributions and weights for these two clusters, indicating different topics.

Table 3.6 also shows some small clusters with documents from a few newsgroups. For instance, Cluster No. 38 has only 31 documents mostly from the three newsgroups, alt.atheism, soc.religion.christian, talk.religion.misc, and even talk.politics.mideast. It indicates a more specific set of news messages. For example, here is an excerpt from a message from the talk.politics.mideast newsgroups that was assigned to Cluster 38 because of the presence of religious words: *"and judgement it is. Until such time as it recognizes that *any* religiously based government is racist, exclusionary and simply built on a philosophy of 'separate but equal' second-class treatment of minorities, it will continue to be known for its bias. If Jewish nationalism is racism, so is Islam; anywhere where people are allotted 'different rights' according to race, religion or culture is 'racist'."*

Some clusters (for instance Cluster No. 0 in Table 3.6) contain documents from almost all newsgroups. Careful examination of some of these documents revealed that most of them do not fit in any of the existing clusters. In fact, their topics are so scattered that they do not form enough of a consensus to form valid clusters. Hence they can be considered as *noise* documents that fall into a *noise magnet* cluster that attracts all noise documents that are not strongly typical of any of the other good clusters. These are documents that lie far away or barely on the border of other clusters (see Section 3.4). In fact Table 3.7 shows that the top 10 relevant keywords

Cluster	1	2	3	4	5	6	7	8	9	10	11	12	13	14	15	16	17	18	19	20
0	2	2	8	12	8	4	5	14	8	6	4	6	7	8	3	1	1	1	11	9
1	14	0	0	0	0	1	1	1	0	0	1	0	0	0	0	2	0	1	0	4
2	3	0	1	1	1	1	1	1	1	3	0	4	1	0	0	3	5	3	4	1
3	3	26	7	11	11	6	1	0	7	1	2	0	8	6	3	7	2	0	0	7
4	2	1	10	0	1	5	1	0	0	1	4	4	1	3	2	0	0	2	2	3
5	1	0	3	1	3	1	1	3	6	0	0	1	1	6	3	4	0	1	1	1
6	9	2	3	3	3	17	1	2	0	1	0	2	4	0	1	0	1	2	1	8
7	1	0	5	12	4	1	4	0	1	2	0	1	1	1	1	1	1	3	1	0
8	1	2	1	1	2	1	8	0	1	0	0	3	8	2	5	0	1	0	0	2
9	0	1	0	5	9	0	3	0	0	0	0	0	4	0	0	1	0	0	0	0
10	0	6	2	3	4	2	1	2	4	1	0	15	1	1	1	2	3	3	1	0
11	3	1	1	0	0	2	3	3	3	12	5	1	0	5	2	3	2	4	4	2
12	0	4	1	1	9	2	5	1	0	0	1	1	8	2	1	0	6	0	0	0
13	6	2	0	2	1	8	5	1	1	4	0	0	0	1	1	1	1	1	0	0
14	1	2	2	4	0	0	2	4	2	2	1	2	2	2	3	0	4	6	4	4
15	0	2	4	1	0	0	1	4	1	0	2	1	0	0	2	21	0	0	0	2
16	0	0	1	0	0	0	1	9	2	0	0	2	6	0	4	1	11	0	6	4
17	1	0	0	1	0	1	1	4	8	6	1	1	0	3	0	4	3	1	3	2
18	3	1	1	0	0	0	0	3	2	1	0	1	0	2	1	2	3	4	1	6
19	1	0	2	0	0	0	3	1	1	4	6	2	2	0	2	1	2	0	1	1
20	0	2	2	0	1	0	0	2	4	8	11	1	1	1	1	0	1	2	3	1
21	2	0	0	1	1	0	0	1	1	0	16	1	0	3	0	1	1	1	1	1
22	1	0	0	2	1	5	4	1	1	1	4	4	6	0	0	0	0	1	1	1
23	1	0	1	0	1	0	0	0	2	2	0	5	3	1	1	1	1	0	16	0
24	3	5	3	2	3	1	3	1	5	3	1	3	0	3	2	2	1	1	1	3
25	3	0	0	0	0	0	0	0	1	0	1	0	2	0	0	2	0	25	0	0
26	0	1	1	2	1	2	0	2	1	4	10	1	0	15	5	0	1	8	2	1
27	3	0	1	0	3	4	2	0	0	2	0	1	0	13	3	2	3	2	1	3
28	1	0	3	0	0	0	3	3	1	0	0	0	0	0	1	0	11	1	8	2
29	2	0	4	0	0	0	2	1	7	0	2	1	0	0	11	0	0	0	1	1
30	1	3	2	0	4	1	1	0	2	4	1	1	1	3	6	1	6	0	2	1
31	1	2	1	1	1	3	0	3	2	3	2	5	1	0	2	4	5	3	2	2
32	3	3	3	1	1	1	0	6	1	9	4	1	0	0	3	0	5	3	1	3
33	1	5	0	0	0	2	3	2	0	0	1	1	0	1	7	2	3	2	4	5
34	5	0	3	1	2	2	2	3	2	0	1	7	1	3	2	1	0	1	1	0
35	1	3	2	1	1	1	1	1	0	2	4	2	2	1	1	7	1	4	2	2
36	0	3	3	9	3	1	1	0	0	1	4	1	2	0	5	3	0	1	1	0
37	1	1	0	1	3	3	2	5	9	3	0	4	2	2	3	0	2	1	5	1
38	7	0	1	0	0	0	0	0	0	2	0	1	0	1	0	6	1	2	0	10
39	1	1	2	1	5	0	2	0	2	0	0	1	2	7	2	1	3	1	1	1

Table 3.6. SKWIC Results: Distribution of the Mini Newsgroup Documents from the 40 Clusters into 20 Prelabeled Classes

have *equally low* relevance weights. In general, the keywords, paired with their relevance weights can be used to infer an automatic (unsupervised) labeling of document clusters.

Finally, we note that some documents are grouped together based solely on commonality of their keyword frequencies. The bag-of-words model is known not to capture the semantics of text. It does not distinguish between different contexts sufficiently to be able to infer that even the same keyword may bear a different meaning. However, this model is much less costly than alternative approaches based on Latent Semantic Indexing (LSI) which may be prohibitively costly for huge, dynamic text collections.

Simulation Results with Fuzzy SKWIC

Table 3.8 shows the class distribution of the 40 clusters discovered by Fuzzy SKWIC, with the columns corresponding to the class indices with complete descriptions listed in Table 3.5. Table 3.9 displays the cluster fuzzy cardinalities ($\sum_{j=1}^{N} \mu_{ij}$), as well as the top 10 relevant keywords for each cluster, sorted in decreasing order of their relevance weights in each cluster.

Table 3.8 shows a more homogeneous class distribution per cluster, indicating a fewer number of documents that risk getting misplaced in a cluster just because they

Cluster	Card	Relevant Words
0	120	abort(0.1127); ford(0.0889); ec(0.0745); matt(0.0684); desktop(0.0638); coverag(0.0625); gordon(0.0554); backup(0.0476); er(0.0387); hill(0.0340);
1	25	atheism(0.8042); trap(0.0653); wisc(0.0228); smart(0.0157); protest(0.0071); dedic(0.0045); ownership(0.0038); absurd(0.0036); arriv(0.0036); probabl(0.0035);
2	34	senat(0.3564); dozen(0.2000); upset(0.1287); corrupt(0.1211); newspap(0.0164); motor(0.0161); remind(0.0076); loui(0.0067); weird(0.0062); pair(0.0057);
3	108	cpu(0.1799); gif(0.1609); ct(0.0983); intel(0.0678); semi(0.0590); geneva(0.0541); tu(0.0517); adob(0.0421); sharewar(0.0294); ch(0.0277);
4	42	app(0.3982); marc(0.1896); fortun(0.1554); ottawa(0.1136); sequenc(0.0179); invent(0.0096); survei(0.0068); forev(0.0053); ration(0.0051); visibl(0.0050);
5	37	nec(0.3993); babi(0.1853); johnson(0.1279); plate(0.0616); radiat(0.0359); hang(0.0272); panel(0.0216); complaint(0.0151); intens(0.0131); ladi(0.0088);
6	60	motif(0.3176); mathew(0.2737); byte(0.1284); grab(0.0719); satisfi(0.0276); entri(0.0202); minim(0.0134); soldier(0.0113); button(0.0110); dedic(0.0082);
7	40	jumper(0.3723); mm(0.2700); sea(0.1036); label(0.0795); quantum(0.0586); aa(0.0279); interrupt(0.0062); er(0.0049); tube(0.0038); avail(0.0038);
8	38	batteri(0.2963); modul(0.1916); blank(0.1426); filter(0.0753); astronomi(0.0501); intens(0.0425); phase(0.0334); accus(0.0214); tune(0.0166); analog(0.0096);
9	23	simm(0.9034); depth(0.0444); phil(0.0074); slot(0.0043); panel(0.0040); macintosh(0.0030); horizont(0.0030); dale(0.0023); hill(0.0022); sea(0.0020);
10	52	privaci(0.2980); databas(0.1988); slot(0.0938); confer(0.0464); mc(0.0392); ration(0.0329); quiet(0.0265); angl(0.0259); pd(0.0257); caught(0.0165);
16	47	atf(0.3067); detector(0.2603); cop(0.1552); radar(0.1205); laser(0.0462); duti(0.0202); border(0.0069); broke(0.0054); trap(0.0053); tear(0.0039);
17	40	cornell(0.3576); pm(0.2517); philosophi(0.0956); shaft(0.0806); cloth(0.0398); england(0.0235); fee(0.0212); drink(0.0083); crew(0.0073); ident(0.0067);
18	31	counter(0.4446); gospel(0.1541); drink(0.0966); deliber(0.0668); disput(0.0474); stretch(0.0444); excess(0.0073); impact(0.0061); tear(0.0061); bias(0.0061);
19	29	miller(0.4506); detroit(0.2479); diego(0.1402); francisco(0.0317); loui(0.0209); bai(0.0078); walker(0.0073); harri(0.0058); psychologi(0.0057); split(0.0046);
20	41	penalti(0.2904); cap(0.1896); worst(0.1142); prism(0.1005); saturdai(0.0766); impact(0.0586); uh(0.0269); fourth(0.0200); capit(0.0096); circumst(0.0065);
21	31	leaf(0.6507); buffalo(0.1242); battl(0.1056); laugh(0.0164); bright(0.0083); ot(0.0067); sad(0.0064); bai(0.0060); hawk(0.0055); pen(0.0045);
22	33	keyboard(0.5893); pen(0.1421); pgp(0.0754); transform(0.0353); lawyer(0.0246); experienc(0.0213); divid(0.0188); england(0.0107); macintosh(0.0061); clone(0.0049);
23	35	cramer(0.3258); clayton(0.1424); accuraci(0.1346); optilink(0.1327); gai(0.0815); survei(0.0587); male(0.0449); mutual(0.0131); bi(0.0049); craig(0.0044);
24	47	msu(0.3999); parallel(0.1671); premis(0.0667); corner(0.0530); onlin(0.0386); exclus(0.0328); cooper(0.0322); bound(0.0293); pixel(0.0262); floor(0.0216);
25	34	armenian(0.4627); turk(0.1566); armenia(0.1145); turkei(0.0949); villag(0.0473); plane(0.0236); border(0.0207); extermin(0.0115); civilian(0.0089); soldier(0.0085);
26	57	sick(0.3158); diet(0.2141); dick(0.1276); graduat(0.1114); huh(0.0391); roughli(0.0215); muscl(0.0149); harder(0.0112); reserv(0.0087); decent(0.0079);
27	43	symptom(0.2385); clue(0.1545); psychologi(0.1261); deriv(0.0902); magic(0.0572); med(0.0454); sad(0.0331); core(0.0288); hide(0.0286); notion(0.0158);
28	34	assault(0.5312); packet(0.1914); followup(0.1068); influenc(0.0619); emerg(0.0067); sentenc(0.0060); exercis(0.0052); girl(0.0047); evil(0.0047); promot(0.0042);
29	32	digex(0.7346); hawk(0.1200); seat(0.0765); joseph(0.0041); intens(0.0031); gear(0.0029); scratch(0.0027); rear(0.0025); carry(0.0025); motor(0.0023);
30	40	planet(0.3477); editor(0.2376); chemic(0.1800); calcul(0.0523); journal(0.0451); newspap(0.0101); atmospher(0.0086); francisco(0.0058); seat(0.0055); mc(0.0043);
31	43	univ(0.2037); anymor(0.1920); walker(0.1360); centr(0.0991); va(0.0912); ridicul(0.0786); crack(0.0185); numer(0.0180); shaft(0.0085); rotat(0.0080);
32	48	superior(0.1537); craig(0.1310); injuri(0.1250); prison(0.1053); incorrect(0.0796); ideal(0.0607); era(0.0486); silver(0.0433); punish(0.0251); string(0.0239);
33	39	purdu(0.4589); apollo(0.2807); solar(0.0791); attornei(0.0461); broke(0.0350); destruct(0.0093); lawyer(0.0076); pair(0.0040); probabl(0.0037); declar(0.0036);
34	37	dont(0.5735); session(0.2218); attract(0.0490); billion(0.0470); worship(0.0219); prism(0.0057); sharewar(0.0050); desktop(0.0037); med(0.0037); implement(0.0032);
35	39	pp(0.3959); credit(0.2320); relationship(0.1024); implement(0.0358); shown(0.0356); clh(0.0235); vers(0.0184); advis(0.0151); graduat(0.0131); declar(0.0074);
36	38	gatewai(0.4673); mon(0.2335); jpl(0.0961); bi(0.0624); phil(0.0371); interrupt(0.0102); buck(0.0060); experienc(0.0045); pitt(0.0042); utexa(0.0038);
37	48	oil(0.1880); alot(0.1690); bag(0.1136); blind(0.1120); weird(0.0957); pair(0.0420); environment(0.0277); eh(0.0272); engag(0.0213); neat(0.0203);
38	31	dwyer(0.2897); judgem(0.2469); horu(0.1252); mchp(0.1252); sni(0.1252); greatest(0.0269); infinit(0.0042); punish(0.0032); walker(0.0032); crack(0.0023);
39	33	iastat(0.4820); tast(0.2779); instrum(0.0955); sector(0.0446); intel(0.0067); filter(0.0061); cloth(0.0028); shield(0.0027); sequenc(0.0026); profit(0.0026);

Table 3.7. SKWIC Results: Cardinality and Term Relevance for the Top 10 Relevant Words in Selected Clusters

lie on areas of overlap. This is because fuzzy memberships develop the partition in a softer and more gradual manner, and hence avoid the early commitment of documents to a specific cluster that occurs with hard 0 or 1 memberships. In fact, it is easier to recognize several meaningful clusters in Table 3.8 with a generally larger number of documents from the same newsgroup, and verify that their relevant keywords, in Table 3.9, are more consistent with the newsgroup's nature than corresponding crisp clusters in Table 3.6. For example, compare the *sci.medical* cluster's (No. 27 in both tables) relevant keywords. Other clusters that are easy to delineate include the two atheism clusters (Nos. 1 and 6), the politics.guns cluster (No. 18), the politics.misc cluster (No. 23), the politics.mideast cluster (No. 25), and the religion.christian cluster (No. 35).

Soft memberships allow a document to belong to several clusters simultaneously, and hence provide a richer model in the areas of overlap. We do not show examples in this section, since we have already illustrated how Fuzzy SKWIC succeeds in providing richer soft labeling for the Web documents in Section 3.5.1. What is worth mentioning in the fuzzy case is that as a result of assigning soft membership degrees to the documents in each cluster, the noise documents, which are roughly equally far from the majority of good clusters, get assigned similar soft memberships in all clusters. Hence they are discouraged from *conspiring* against one of the clusters as in the crisp partitioning framework, where they can acquire a *whole* membership of 1 in a given cluster because of arbitrary crisp assignment based on minimum (within ϵ) distance. This means that, generally, noise documents will have almost equal memberships $(1/C)$ in all clusters, hence their influence on good clusters is broken up into smaller equal pieces instead of a whole sum. Consequently, their net effect on the resulting partition and all estimated parameters (since everything is weighted by the memberships) gets diluted, and this is what makes our *soft* partitioning strategy more *robust* to noise. A direct consequence of this fact is that there is no longer a big noise cluster grouping several documents from all newsgroups as in the crisp case (Cluster No. 3).

We note that despite the *softness* of the memberships, the clusters that are very homogeneous in the nature of their documents end up with almost *crisp* 0 - 1 memberships. Hence the crisp partition is a special case of soft partitioning that does emerge when there is no strong overlap between different clusters.

We have further performed clustering using unweighted keyword-based techniques: K Means and the Fuzzy C Means, (both with cosine-based distance) and have noticed that both crisp and fuzzy SKWIC tend to outperform their unweighted counterparts. For instance, the noise cluster that grabs documents from all different newsgroups gets even larger. To summarize, K Means lies on the least favorable side of the spectrum because it has no way of adapting different clusters to capture different relevance degrees in their keywords, nor different membership degrees of their documents. SKWIC is able to model different keyword relevance degrees depending on the cluster, but cannot model gradual degrees of membership of documents. The Fuzzy C Means fails to model different cluster-dependent keyword relevance degrees but can model gradual degrees of membership of documents. Hence both SKWIC and Fuzzy C Means have complementary but exclusive strengths that

Cluster	Classes																			
	1	2	3	4	5	6	7	8	9	10	11	12	13	14	15	16	17	18	19	20
0	4	3	2	3	2	23	5	0	1	1	0	1	0	1	1	1	0	0	0	1
1	14	0	0	0	0	3	1	0	0	0	0	1	0	0	0	1	1	0	0	4
2	3	2	5	8	2	0	0	0	5	3	2	3	1	1	2	3	1	1	1	1
3	0	18	3	2	5	1	1	0	4	0	0	1	2	1	1	0	0	0	0	1
4	1	1	9	1	1	4	1	2	2	4	0	0	1	1	0	1	0	0	1	0
5	4	0	2	1	2	1	3	2	7	0	0	1	4	1	0	1	1	2	1	1
6	10	2	0	1	0	1	0	0	1	2	0	0	0	0	0	1	0	0	2	6
7	1	1	4	13	7	0	1	1	0	2	0	3	1	0	0	0	0	4	0	0
8	6	2	7	1	3	0	0	1	3	2	0	14	3	4	4	2	4	0	0	5
9	0	0	0	5	10	0	2	0	0	0	0	0	4	0	0	0	0	0	0	0
10	0	5	0	0	0	2	2	0	2	2	2	9	0	1	0	0	2	3	2	1
11	2	0	1	7	0	3	2	2	4	4	0	4	9	1	1	3	3	3	6	
12	0	4	1	1	6	2	9	1	1	0	1	1	9	3	1	0	3	1	1	0
13	3	1	3	1	1	1	1	1	1	7	0	5	4	1	1	0	1	0	0	2
14	2	2	2	6	2	2	5	17	6	4	3	0	3	1	8	0	1	2	3	1
15	2	2	3	3	3	0	1	4	3	0	3	2	0	0	1	16	2	0	0	4
16	0	1	0	0	2	0	2	7	3	0	0	0	7	0	6	0	3	0	2	0
17	0	0	0	3	1	2	1	5	9	7	1	0	0	4	1	3	2	1	4	2
18	1	1	3	1	0	1	1	1	0	1	0	2	0	1	0	1	21	1	8	7
19	1	2	2	0	1	3	1	2	2	7	19	1	1	3	4	1	4	4	5	0
20	0	1	3	2	6	0	3	2	1	4	8	1	4	0	1	0	0	0	2	1
21	0	0	0	0	0	0	0	1	0	0	14	0	0	3	0	0	1	0	0	0
22	0	0	3	7	10	6	6	0	1	3	0	3	8	1	0	0	1	0	0	2
23	1	0	2	0	0	0	0	2	0	2	3	1	1	0	3	0	1	0	17	0
24	1	0	2	1	4	1	2	5	6	2	2	1	2	1	3	2	3	1	2	2
25	2	4	0	0	0	3	0	0	0	0	1	0	4	0	1	1	0	30	0	0
26	3	1	1	1	0	1	0	2	0	5	6	1	1	6	3	1	3	6	3	3
27	0	0	0	0	0	1	0	1	0	0	2	4	0	0	34	1	0	1	0	0
28	0	1	0	4	0	3	2	4	1	4	1	5	0	1	4	5	5	2	10	5
29	1	0	4	0	0	0	0	1	0	0	0	1	0	1	11	0	0	0	1	1
30	1	3	1	0	0	1	4	5	9	3	6	1	2	1	4	1	3	1	1	0
31	3	2	0	1	1	3	0	2	5	4	2	3	1	4	3	5	6	1	5	3
32	4	2	2	1	0	1	2	2	2	5	1	4	1	0	4	5	6	6	5	7
33	1	3	0	0	0	2	3	2	0	0	1	2	0	0	4	2	3	5	5	6
34	4	2	6	2	1	4	0	2	4	1	1	13	0	4	1	2	1	2	2	1
35	2	1	2	1	1	2	2	1	2	1	2	2	1	2	0	25	1	8	0	3
36	2	5	3	1	0	1	4	1	0	1	0	1	3	0	13	1	2	0	1	1
37	1	8	3	0	6	1	4	5	4	3	1	4	2	1	3	0	4	5	5	4
38	7	0	1	0	0	0	0	0	0	2	0	1	0	1	0	5	1	1	0	10
39	1	1	4	2	9	0	2	1	2	0	0	1	5	1	1	0	2	1	1	3

Table 3.8. Fuzzy SKWIC Results: Distribution of the Mini Newsgroup Documents from 40 Clusters into 20 Prelabeled Classes

make them provide richer partition models. However, Fuzzy SKWIC lies on the most favorable side of the spectrum because it is able to provide both dynamic soft degrees in the keyword relevance values and in the cluster memberships, and can be thus considered to perform simultaneous partitioning in two different hyperspaces: the document space to capture *spatial* document organization, and the keyword space to capture *context*. The context can be inferred for each cluster because it is described in terms of several relevant keywords, and these keywords are further given importance degrees that vary with each cluster. The context stems mainly from the well-known fact that it is easier to infer context from *several* keywords simultaneously, than from any single one of the keywords. The relevance weights are expected to further enrich the context description.

Simulation Results on the 16 Mini Newsgroup Data and Entire 20 Newsgroups Data

With the Mini Newsgroup data set in the previous section, we have noticed that there were several misclassified (or inconsistently assigned) documents that come from the four miscellaneous classes (Nos. 3, 7, 19, 20). Most of these documents have been assumed to have the same groundtruth label (newsgroup name), but

Cluster	Card	Relevant Words
0	43.87	motif(0.5799); default(0.1127); string(0.0882); height(0.0460); depth(0.0448); hang(0.0234); focu(0.0103); button(0.0081); databas(0.0053); byte(0.0052);
1	25.52	atheism(0.9049); wisc(0.0078); trap(0.0049); ownership(0.0043); arriv(0.0042); absurd(0.0041); probabl(0.0040); dedic(0.0040); suitabl(0.0039); declar(0.0036);
2	41.52	gatewai(0.5335); upset(0.1644); mirror(0.0621); bi(0.0606); corrupt(0.0484); phil(0.0301); suck(0.0048); utexa(0.0048); batteri(0.0045); suddenli(0.0044);
3	42.36	gif(0.4690); pixel(0.1592); slot(0.1216); clip(0.1017); pd(0.0606); blank(0.0084); implement(0.0071); domain(0.0058); plane(0.0047); sharewar(0.0045);
4	33.55	app(0.5580); lee(0.2951); decent(0.0500); favorit(0.0040); staff(0.0040); intens(0.0038); tech(0.0037); superior(0.0031); sea(0.0027); bag(0.0026);
5	30.58	nec(0.5792); babi(0.2441); lawyer(0.0545); ladi(0.0111); dog(0.0093); armi(0.0075); odd(0.0064); punish(0.0050); turk(0.0046); joseph(0.0045);
6	29.40	mathew(0.7106); ideal(0.1854); quantum(0.0099); resembl(0.0067); turkei(0.0051); civilian(0.0051); dwyer(0.0041); decent(0.0039); interrupt(0.0037); greatest(0.0036);
7	33.67	jumper(0.5789); quantum(0.2108); interrupt(0.0914); advoc(0.0113); corrupt(0.0079); movi(0.0065); avail(0.0056); label(0.0055); speech(0.0050); convent(0.0050);
8	53.64	privaci(0.2250); fortun(0.1716); carl(0.1527); premis(0.1092); modul(0.0939); perman(0.0411); exclud(0.0297); swap(0.0256); blank(0.0239); habit(0.0060);
9	26.37	simm(0.9450); phil(0.0077); panel(0.0046); slot(0.0045); depth(0.0037); macintosh(0.0031); horizont(0.0031); dale(0.0027); sea(0.0023); wise(0.0017);
10	36.54	databas(0.3113); senat(0.2615); confer(0.2356); caught(0.0332); foreign(0.0284); pa(0.0105); punish(0.0049); fourth(0.0047); forev(0.0046); crack(0.0045);
16	32.48	detector(0.4793); radar(0.2463); cop(0.1577); duti(0.0265); border(0.0128); laser(0.0124); worker(0.0072); angl(0.0063); trap(0.0051); max(0.0047);
17	42.61	cornell(0.3754); pm(0.2749); drink(0.1067); shaft(0.1004); loui(0.0096); wound(0.0064); austin(0.0055); fee(0.0054); utexa(0.0054); attend(0.0050);
18	54.42	atf(0.3720); assault(0.2372); packet(0.1930); stretch(0.0427); pointer(0.0405); camera(0.0071); threat(0.0059); broke(0.0059); lawyer(0.0058); attornei(0.0047);
19	57.56	pp(0.2393); penalti(0.2060); chemic(0.1558); impact(0.0866); detroit(0.0789); prism(0.0737); worst(0.0538); capit(0.0092); loui(0.0061); circl(0.0056);
20	41.42	backup(0.5167); cap(0.2153); pen(0.0721); laser(0.0657); analog(0.0338); fourth(0.0051); classic(0.0043); devil(0.0038); attend(0.0035); buffalo(0.0029);
21	24.31	leaf(0.8989); bai(0.0083); hawk(0.0076); pen(0.0062); detroit(0.0060); devil(0.0057); leg(0.0055); bright(0.0051); lee(0.0051); carl(0.0039);
22	51.57	cpu(0.4203); keyboard(0.3074); catch(0.0928); macintosh(0.0710); transform(0.0286); clone(0.0097); suitabl(0.0055); speaker(0.0055); blind(0.0053); tech(0.0045);
23	34.45	cramer(0.3586); clayton(0.1669); survei(0.1506); optilink(0.1460); gai(0.0765); mutual(0.0277); male(0.0142); bi(0.0054); ottawa(0.0052); sea(0.0042);
24	44.52	msu(0.4406); chain(0.2240); devil(0.0981); visibl(0.0609); exclus(0.0564); trap(0.0106); negoti(0.0058); oil(0.0043); effici(0.0043); session(0.0042);
25	51.48	armenian(0.3693); tu(0.1333); soldier(0.0877); armenia(0.0873); turk(0.0715); turkei(0.0559); villag(0.0363); border(0.0338); extermin(0.0096);
26	41.72	sick(0.5233); counter(0.1695); huh(0.0986); roughli(0.0467); disput(0.0449); bias(0.0113); accus(0.0084); amateur(0.0063); harder(0.0052); walker(0.0047);
27	49.52	diet(0.3122); symptom(0.2068); tast(0.1057); med(0.0869); literatur(0.0442); muscl(0.0411); healthi(0.0392); clinic(0.0134); root(0.0123); mouth(0.0116);
28	57.65	abort(0.3024); matt(0.2011); coverag(0.1483); workstat(0.1115); followup(0.0667); denni(0.0476); protest(0.0078); acknowledg(0.0064); emerg(0.0063); convict(0.0056);
29	26.36	digex(0.9382); intens(0.0041); gear(0.0037); scratch(0.0036); carry(0.0033); motor(0.0032); bound(0.0027); restor(0.0026); carl(0.0025); confer(0.0023);
30	44.66	craig(0.2987); seat(0.2141); editor(0.1603); hawk(0.1180); francisco(0.0753); floor(0.0089); batteri(0.0088); mid(0.0083); rear(0.0067); candid(0.0044);
31	47.69	adob(0.2839); univ(0.2235); anymor(0.1875); johnson(0.0946); laugh(0.0386); va(0.0242); evil(0.0107); exercis(0.0079); weird(0.0059); pa(0.0059);
32	40.30	battl(0.3249); superior(0.2138); hide(0.1514); fee(0.0549); tear(0.0549); dy(0.0221); root(0.0085); deliber(0.0080); forev(0.0078); glad(0.0070);
33	42.43	purdu(0.4496); apollo(0.3028); attornei(0.0690); prison(0.0597); broke(0.0208); destruct(0.0076); lawyer(0.0073); pair(0.0053); probabl(0.0041); declar(0.0039);
34	51.53	dont(0.4483); session(0.1581); pgp(0.1095); england(0.0695); implement(0.0609); worship(0.0521); prism(0.0050); entri(0.0049); desktop(0.0040); australia(0.0039);
35	51.93	geneva(0.3260); graduat(0.1728); male(0.0931); relationship(0.0915); vers(0.0658); clh(0.0561); harri(0.0260); credit(0.0148); adult(0.0106); proven(0.0104);
36	38.55	miller(0.3295); sharewar(0.2967); jpl(0.1575); solar(0.1126); fee(0.0075); marc(0.0052); convent(0.0052); sea(0.0051); psychologi(0.0050); attract(0.0042);
37	62.63	ch(0.2326); su(0.1411); domain(0.0870); planet(0.0779); negoti(0.0597); ownership(0.0579); advertis(0.0549); environment(0.0459); weird(0.0415); marc(0.0403);
38	33.43	dwyer(0.3045); judgem(0.2431); horu(0.1317); mchp(0.1317); sni(0.1317); infinit(0.0045); greatest(0.0039); walker(0.0038); punish(0.0036); crack(0.0025);
39	35.46	iastat(0.4500); intel(0.4329); sector(0.0417); instrum(0.0055); battl(0.0048); forev(0.0043); tast(0.0043); optic(0.0027); shield(0.0026); button(0.0026);

Table 3.9. Fuzzy SKWIC Results: Fuzzy Cardinality and Term Relevance for the Top 10 Relevant Words in Selected Clusters

their contents vary widely in topic, in a way that would make some of them more appropriately labeled with other newsgroup names. Therefore, we repeated all experiments after discarding documents from these four classes. This means that we removed most of the *difficult* cases, so to speak. These include, (i) documents that lie in areas of overlap or fuzziness between distinct categories, or (ii) documents that are simply outliers, and hence affect the purity of the resulting partition. After discarding the .misc classes, we noticed similar results in terms of the nature of the clusters, and the richer information provided by the cluster-dependent keyword relevance weights, and soft partition. We also noticed a remarkable improvement in the purity of the partition with regard to cluster homogeneity, as compared to including the miscellaneous class documents.

One way to objectively assess the performance of a clustering algorithm when the class labels for K classes are actually known, is based on the average entropy measure of all C clusters, which is defined as follows.

$$E = \sum_{i=1}^{C} \frac{N_i}{N} E_i,$$

where

$$E_i = \frac{1}{\log K} \sum_{k=1}^{K} \frac{N_i^k}{N_i} \log \frac{N_i^k}{N_i},$$

is the entropy of the ith cluster, N_i is the size of the ith cluster, and N_i^k is the number of documents from the kth class that are assigned to the ith cluster. Table 3.10 lists the entropies of the partitioning strategies used for the Mini and 20 Newsgroup data, depending on whether the four miscellaneous classes are removed.

With all the empirical results and theoretically based conclusions about the data sets used in this chapter in mind, the most important fact to remember is that in our *nonideal* world, *real unlabeled* text data tend to be of the *challenging type* discussed above. This in turn calls for sophisticated techniques that can handle these challenges.

We also note that with the 20 Newsgroups data set, as with almost any manually labeled benchmark document data set, errors in labeling abound (due to subjectivity in labeling, circumstantial reasons, or even noise documents that still end up with an invalid label). Also, documents that cross boundaries between different categories are very common, and always end up with an inadequate label. Hence, it is extremely difficult to judge the quality of an unsupervised clustering technique based on any kind of classification accuracy or *entropy* measure. In fact, our experiments have shown that automatic labeling is often superior to manual labeling, except when identical keywords with different meanings are present. This is where keyword-based clustering breaks down because it cannot deal with the semantics of content. For such cases, context can improve clustering results considerably, and this can be handled using Latent Semantic Indexing [DDF+90, BDJ99], for example.

	Mini Newsgroups	Mini Newsgroups 16 Class	20 Newsgroups
K Means	0.797	0.771	0.865
SKWIC	0.790	0.750	0.866
Fuzzy C Means	0.766	0.751	0.907
Fuzzy SKWIC	0.757	0.740	0.868

Table 3.10. Average Entropies for Different Categorization Strategies of the Newsgroup Data

3.6   Conclusion

In this chapter, we presented a new approach that performs clustering and attribute weighting simultaneously and in an unsupervised manner. Our approach is an extension of the K-Means algorithm, that in addition to partitioning the data set into a given number of clusters, also finds an optimal set of feature weights for *each* cluster. SKWIC minimizes one objective function for both the optimal prototype parameters and feature weights for each cluster. This optimization is done iteratively by dynamically updating the prototype parameters and the attribute relevance weights in each iteration. This makes the proposed algorithm simple and fast.

Our experimental results showed that SKWIC outperforms K-Means when not all the features are equally relevant to all clusters. This makes our approach more reliable, especially when clustering in *high-dimensional* spaces, as in the case of text document categorization, where not all attributes are equally important, and where clusters tend to form in only *subspaces* of the original feature space. Also, for the case of *text* data, this approach can be used to automatically annotate the documents.

We have also presented a soft partitioning approach (Fuzzy SKWIC) to handle the inherent fuzziness in text documents by automatically generating fuzzy or soft labels instead of single-label categorization. This means that a text document can belong to *several* categories with different degrees. The soft approach succeeds in describing documents at the intersection between several categories.

By virtue of the dynamic keyword weighting, and its continuous interaction with distance and membership computations, the proposed approach is able to handle noise documents elegantly by automatically designating one or two *noise magnet* clusters that grab most outliers away from the other clusters.

Compared to variants such as K-Means and the Fuzzy C-Means, Fuzzy SKWIC is able to provide both dynamic soft degrees in the keyword relevance values and in the cluster memberships, and can be thus considered to perform simultaneous partitioning in two different hyperspaces: the document space to capture *spatial* document organization, and the keyword space to capture *context*. The context stems mainly from the well-known fact that it is easier to infer context from a consensus of *several* keywords simultaneously, than from any single one of the keywords. The relevance weights are expected to further enrich the context description.

Our results have also confirmed that Fuzzy SKWIC is most appropriate to use in cases where the document collection is *challenging*, meaning that it may be limited in terms of the number of documents, and the number of keywords used to infer the labels, and that it may include many *noise* documents and *mixed-topic* documents that blur the boundaries between clusters. Our *nonideal* world abounds with *unlabeled* text data that tend to be of the *challenging type*. Fuzzy SKWIC is one of the unsupervised classification techniques that can handle these challenges.

Since the objective function of SKWIC is based on that of the K-Means, it inherits most of the advantages of K-Means-type clustering algorithms, such as ease of computation and simplicity. Moreover, because K-Means has been studied extensively over the last decades, the proposed approach can easily benefit from the advances and improvements that led to several K-Means variants in the data mining and pattern recognition communities; in particular, the techniques developed to handle noise [KK93, NK96, FK99, NK97], to determine the number of clusters [FK97], to cluster very large data sets [BFR98, FLE00], and to improve initialization [BF98, HOB99, NK00]. Future research includes investigating more scalable extensions that are not sensitive to initialization, and that can determine the optimal number of clusters. We also plan to explore context-dependent information retrieval based on a combination of Fuzzy SKWIC with concepts from fuzzy logic, particularly its ability to *compute with words*.

Acknowledgments: We thank our student Mrudula Pavuluri for her help with preprocessing some of the text data. Partial support of this work was provided by the National Science Foundation through CAREER Awards IIS 0133415 (Hichem Frigui) and IIS 0133948 (Olfa Nasraoui).

References

[AD91]   H. Almuallim and T.G. Dietterich.Learning with many irrelevant features.In *Proceedings of the Ninth National Conference on Artificial Intelligence*, pages 547–552, 1991.

[BDJ99]   M. Berry, Z. Drmač, and E. Jessup.Matrices, vector spaces, and information retrieval.*SIAM Review*, 41(2):335–362, 1999.

[Bez81]   J.C. Bezdek.*Pattern Recognition with Fuzzy Objective Function Algorithms*.Plenum, New York, 1981.

[BF98]   P.S. Bradley and U.M. Fayyad.Refining initial points for K-Means clustering.In *Procedings of the Fifteenth International Conference on Machine Learning*, Morgan Kaufmann, San Francisco, pages 91–99, 1998.

[BFR98]   P.S. Bradley, U.M. Fayyad, and C. Reina.Scaling clustering algorithms to large databases.In *Knowledge Discovery and Data Mining*, pages 9–15, 1998.

[BL85]   C. Buckley and A.F. Lewit.Optimizations of inverted vector searches.In *SIGIR '85*, pages 97–110, 1985.

[Bow]   Bow: A toolkit for statistical language modeling, text retrieval, classification and clustering [online, cited September 2002].Available from World Wide Web: www.cs.cmu.edu/~mccallum/bow.

[CKPT92]   D.R. Cutting, D.R. Karger, J.O. Pedersen, and J.W. Turkey.Scatter/gather: A cluster-based approach to browsing large document collections.In *Proceedings of the Fifteenth Annual International ACM SIGIR Conference on Research and Development in Information Retrieval*, Copenhagen, Denmark, pages 318–329, June 1992.

[CMU]   20 newsgroup data set [online, cited September 2002].Available from World Wide Web: www-2.cs.cmu.edu/afs/cs.cmu.edu/project/theo-20/www/data/news20.html.

[DDF$^+$90]   S. Deerwester, S. Dumais, G. Furnas, T. Landauer, and R. Harshman.Indexing by latent semantic analysis.*Journal of the American Society for Information Science*, 41(6):391–407, 1990.

[FK97]   H. Frigui and R. Krishnapuram.Clustering by competitive agglomeration.*Pattern Recognition*, 30(7):1223–1232, 1997.

[FK99]   H. Frigui and R. Krishnapuram.A robust competitive clustering algorithm with applications in computer vision.*IEEE Transactions on Pattern Analysis and Machine Intelligence*, 21(5):450–465, May 1999.

[FLE00]   F. Farnstrom, J. Lewis, and C. Elkan.Scalability for clustering algorithms revisited.*SIGKDD Explorations*, 2(1):51–57, 2000.

[FN00]   H. Frigui and O. Nasraoui.Simultaneous clustering and attribute discrimination.In *Proceedings of the IEEE Conference on Fuzzy Systems*, San Antonio, TX, pages 158–163, 2000.

[GK79]   E.E. Gustafson and W.C. Kessel.Fuzzy clustering with a fuzzy covariance matrix.In *Proceedings of the IEEE Conference on Decision and Control*, San Diego, pages 761–766, 1979.

[HOB99]   L.O. Hall, I.O. Ozyurt, and J.C. Bezdek.Clustering with a genetically optimized approach.*IEEE Transactions on Evolutionary Computations*, 3(2):103–112, Jul 1999.

[Hub81]   P.J. Huber.*Robust Statistics*.Wiley, New York, 1981.

[JKP94]   G. John, R. Kohavi, and K. Pfleger.Irrelevant features and the subset selection problem.In *Proceedings of the Eleventh International Machine Learning Conference*, pages 121–129, 1994.

[KK93]   R. Krishnapuram and J. M. Keller.A possibilistic approach to clustering.*IEEE Transactions on Fuzzy Systems*, 1(2):98–110, May 1993.

[Kor77]   R.R. Korfhage.*Information Storage and Retrieval*.Wiley, New York, 1977.

[Kow97]   G. Kowalski.*Information Retrieval Systems: Theory and Implementation*. Kluwer Academic, Hingham, MA, 1997.

[KR92]   K. Kira and L. A. Rendell.The feature selection problem: Traditional methods and a new algorithm.In *Proceedings of the Tenth National Conference on Artificial Intelligence*, pages 129–134, 1992.

[KS95]   R. Kohavi and D. Sommerfield.Feature subset selection using the wrapper model: Overfitting and dynamic search space topology.In *Proceedings of the First International Conference on Knowledge Discovery and Data Mining*, pages 192–197, 1995.

[Mla99]   D. Mladenic.Text learning and related intelligent agents.*IEEE Expert*, Jul 1999.

[NK96]   O. Nasraoui and R. Krishnapuram.An improved possibilistic c-means algorithm with finite rejection and robust scale estimation.In *Proceedings of the North American Fuzzy Information Processing Society Conference*, Berkeley, CA, pages 395–399, Jun 1996.

[NK97]   O. Nasraoui and R. Krishnapuram.A genetic algorithm for robust clustering based on a fuzzy least median of squares criterion.In *Proceedings of the North American Fuzzy Information Processing Society Conference*, Syracuse, NY, pages 217–221, Sept 1997.

[NK00]   O. Nasraoui and R. Krishnapuram.A novel approach to unsupervised robust clustering using genetic niching.In *Proceedings of the IEEE International Conference on Fuzzy Systems*, New Orleans, pages 170–175, 2000.

[RK92]   L.A. Rendell and K. Kira.A practical approach to feature selection.In *Proceedings of the International Conference on Machine Learning*, pages 249–256, 1992.

[RL87]   P. J. Rousseeuw and A. M. Leroy.*Robust Regression and Outlier Detection*.Wiley, New York, 1987.

[Ska94]   D. Skalak.Prototype and feature selection by sampling and random mutation hill climbing algorithms.In *Proceedings of the Eleventh International Machine Learning Conference (ICML-94)*, pages 293–301, 1994.

[vR79]   C.J. van Rijsbergen.*Information Retrieval second edition*.Butterworths, London, 1979.

[ZEMK97] O. Zamir, O. Etzioni, O. Madani, and R.M. Karp.Fast and intuitive clustering of web documents.In *KDD'97*, pages 287–290, 1997.

4

Feature Selection and Document Clustering

Inderjit Dhillon
Jacob Kogan
Charles Nicholas

Overview

Feature selection is a basic step in the construction of a vector space or *bag-of-words* model [BB99]. In particular, when the processing task is to partition a given document collection into clusters of similar documents a choice of good features along with good clustering algorithms is of paramount importance. This chapter suggests two techniques for feature or term selection along with a number of clustering strategies. The selection techniques significantly reduce the dimension of the vector space model. Examples that illustrate the effectiveness of the proposed algorithms are provided.

4.1 Introduction

A common form of text processing in many information retrieval systems is based on the analysis of word occurrences across a document collection. The number of words/terms used by the system defines the dimension of a vector space in which the analysis is carried out. Reduction of the dimension may lead to significant savings of computer resources and processing time. However, poor feature selection may dramatically degrade the information retrieval system's performance.

Dhillon and Modha [DM01] have recently used the spherical k-means algorithm for clustering text data. In one of the experiments of [DM01] the algorithm was applied to a data set containing 3893 documents. The data set contained the following three document collections (available from `ftp://ftp.cs.cornell.edu/pub/smart`):

- Medlars Collection (1033 medical abstracts),

- CISI Collection (1460 information science abstracts),

- Cranfield Collection (1400 aerodynamics abstracts).

Partitioning the entire collection into three clusters generates the following "confusion" matrix reported in [DM01].

	Medlars	CISI	Cranfield
cluster 0	1004	5	4
cluster 1	18	1440	16
cluster 2	11	15	1380

(Here the entry ij is the number of documents that belong to cluster i and document collection j.) The confusion matrix shows that only 69 documents (i.e., less that 2% of the entire collection) have been "misclassified" by the algorithm. After removing stop-words Dhillon and Modha [DM01] reported 24,574 unique words, and after eliminating low- and high-frequency words they selected 4,099 words to construct the vector space model.

The main goal of this contribution is to provide algorithms for (i) selection of a small set of terms and (ii) clustering of document vectors. In particular, for data similar to those described above, we are able to generate better or similar quality confusion matrices while reducing the dimension of the vector space model by more than 70%.

The outline of the chapter is the following. A brief review of existing algorithms we employ for clustering documents is provided in Section 4.2. The data are described in Section 4.3. The term-selection techniques along with the clustering results are presented in Sections 4.4 and 4.5, while Section 4.6 contains a new clustering algorithm along with the corresponding clustering results. Future research directions are briefly outlined in Section 4.7.

4.2   Clustering Algorithms

In this section, we review two known clustering algorithms that we apply to partition documents into clusters. The means algorithm, introduced in [Kog01b], is a combination of the batch k-means and the incremental k-means algorithms (see [DHS01]). The Principal Direction Divisive Partitioning (PDDP) method was introduced recently by D. Boley [Bol98].

4.2.1   Means Clustering Algorithm

For a set of vectors $\mathbf{X} = \{\mathbf{x}_1, \ldots, \mathbf{x}_d\}$ in Euclidean space $\mathbf{R}^w$ denote the centroid of the set $(1/d) \sum_{i=1}^{d} \mathbf{x}_i$ by $\mathbf{m}(\mathbf{X})$.

Let $\{\pi_l\}_{l=1}^k$ be a partition of $\mathbf{X}$ with the corresponding centroids $\mathbf{m}_1 = \mathbf{m}(\pi_1), \ldots, \mathbf{m}_k = \mathbf{m}(\pi_k)$. Define the quality Q_2 of the partition $\{\pi_l\}_{l=1}^k$ by

$$Q_2\left(\{\pi_l\}_{l=1}^k\right) = \sum_{l=1}^k \sum_{\mathbf{x} \in \pi_l} \|\mathbf{x} - \mathbf{m}(\pi_l)\|^2 = \sum_{l=1}^k \sum_{\mathbf{x} \in \pi_l} \|\mathbf{x} - \mathbf{m}_l\|^2. \tag{4.1}$$

For $\mathbf{x} \in \pi_i \subseteq \mathbf{X}$ denote the index of the centroid nearest $\mathbf{x}$ by $\min(\mathbf{x})$ (i.e., $\|\mathbf{x} - \mathbf{m}_{\min(\mathbf{x})}\| \leq \|\mathbf{x} - \mathbf{m}_l\|$, $l = 1, \ldots, k$). Now define the partition $\mathrm{nextKM}\left(\{\pi_l\}_{l=1}^k\right) = \{\pi_l'\}_{l=1}^k$ as

$$\pi_i' = \{\mathbf{x} : \min(\mathbf{x}) = i\}.$$

It is easy to see [DHS01] that

$$Q_2\left(\{\pi_l\}_{l=1}^k\right) \geq Q_2\left(\mathrm{nextKM}\left(\{\pi_l\}_{l=1}^k\right)\right). \tag{4.2}$$

Next we present the classical batch k-means algorithm and discuss some of its deficiencies.

Algorithm 4.1 Batch k-means clustering algorithm (Forgy [For65])

For a user-supplied tolerance $\mathtt{tol} < 0$ do the following.

1. Start with an arbitrary partitioning $\left\{\pi_l^{(0)}\right\}_{l=1}^k$. Set the index of iteration $t = 0$.

2. Generate the partition $\mathrm{nextKM}\left(\left\{\pi_l^{(t)}\right\}_{l=1}^k\right)$.

   If $\left[Q_2\left(\mathrm{nextKM}\left(\left\{\pi_l^{(t)}\right\}_{l=1}^k\right)\right) - Q_2\left(\left\{\pi_l^{(t)}\right\}_{l=1}^k\right) \leq \mathtt{tol}\right]$

   set $\left\{\pi_l^{(t+1)}\right\}_{l=1}^k = \mathrm{nextKM}\left(\left\{\pi_l^{(t)}\right\}_{l=1}^k\right)$
   increment t by 1.
   go to 2.

3. Stop.

The algorithm suffers from the major drawbacks:

1. The quality of the final partition depends on a good choice of the initial partition; and

2. The algorithm may get trapped at a local minimum even for a very simple one-dimensional set $\mathbf{X}$.

We address the first point in Sections 4.2.2 and 4.6. The second point is illustrated by the following example.

EXAMPLE 4.1 *Let* $\mathbf{X} = \{\mathbf{x}_1, \mathbf{x}_2, \mathbf{x}_3\}$ *with* $\mathbf{x}_1 = 0$, $\mathbf{x}_2 = 2/3$, *and* $\mathbf{x}_3 = 1$. *Consider the initial partition* $\pi_1^{(0)} = \{\mathbf{x}_1, \mathbf{x}_2\}$, $\pi_2^{(0)} = \{\mathbf{x}_3\}$ *with* $\mathcal{Q}_2\left\{\pi_1^{(0)}, \pi_2^{(0)}\right\} = 2/9$.

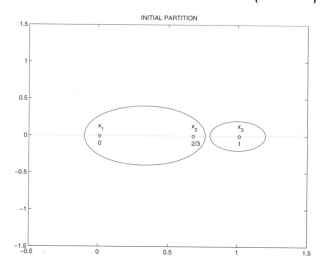

Note that an application of the batch k-means algorithm does not change the initial partition $\left\{\pi_1^{(0)}, \pi_2^{(0)}\right\}$. *At the same time it is clear that the partition* $\{\pi_1', \pi_2'\}$ *($\pi_1' = \{\mathbf{x}_1\}$, $\pi_2' = \{\mathbf{x}_2, \mathbf{x}_3\}$) with* $\mathcal{Q}_2\{\pi_1', \pi_2'\} = 2/36$ *is superior to the initial partition.*

A different version of the k-means algorithm, incremental k-means clustering, is discussed next. This version remedies the problem illustrated in Example 4.1.

The decision of whether a vector $\mathbf{x} \in \pi_i$ should be moved from cluster π_i to cluster π_j is made by the batch k-means algorithm based on the sign of

$$\Delta = - \|\mathbf{x} - \mathbf{m}(\pi_i)\|^2 + \|\mathbf{x} - \mathbf{m}(\pi_j)\|^2. \tag{4.3}$$

If Δ is negative, then the vector $\mathbf{x}$ is moved by the batch k-means algorithm. The exact change in the value of the objective function (i.e., the difference between the "new" and the "old" values of the objective function) caused by the move is

$$\Delta_{\text{exact}} = -\frac{n_i}{n_i - 1} \|\mathbf{x} - \mathbf{m}(\pi_i)\|^2 + \frac{n_j}{n_j + 1} \|\mathbf{x} - \mathbf{m}(\pi_j)\|^2, \tag{4.4}$$

where $n_j = |\pi_j|$, $n_i = |\pi_i|$ are the number of vectors in clusters π_j and π_i, respectively (see, e.g., [Kog01a]). The more negative Δ_{exact} is, the larger the drop in the value of the objective function. The difference between the expressions

$$\Delta - \Delta_{\text{exact}} = \frac{1}{n_i - 1} \|\mathbf{x} - \mathbf{m}(\pi_i)\|^2 + \frac{1}{n_j + 1} \|\mathbf{x} - \mathbf{m}(\pi_j)\|^2 \geq 0$$

is negligible when the clusters π_i and π_j are large. However $\Delta - \Delta_{\text{exact}}$ may become significant for small clusters. For example, for $\mathbf{x} = \mathbf{x}_2$ in Example 4.1

one has $\Delta = 0$, and $\Delta_{\text{exact}} < 0$. This is why batch k-means misses the "better" partition $\{\pi'_1, \pi'_2\}$. The incremental k-means clustering algorithm eliminates this problem. Before presenting the algorithm, we need a few additional definitions.

DEFINITION 4.1 *A first variation of a partition* $\{\pi_l\}_{l=1}^{k}$ *is a partition* $\{\pi'_l\}_{l=1}^{k}$ *obtained from* $\{\pi_l\}_{l=1}^{k}$ *by removing a single vector* $\mathbf{x}$ *from a cluster* π_i *of* $\{\pi_l\}_{l=1}^{k}$ *and assigning this vector to an existing cluster* π_j *of* $\{\pi_l\}_{l=1}^{k}$.

Note that the partition $\{\pi_l\}_{l=1}^{k}$ is a first variation of itself. Next we look for the *steepest descent* first variation, that is, a first variation that leads to the maximal decrease of the objective function. The formal definition follows.

DEFINITION 4.2 *The partition* $\mathtt{nextFV}\left(\{\pi_l\}_{l=1}^{k}\right)$ *is a first variation of* $\{\pi_l\}_{l=1}^{k}$ *so that for each first variation* $\{\pi'_l\}_{l=1}^{k}$ *one has*

$$Q_2\left(\mathtt{nextFV}\left(\{\pi_l\}_{l=1}^{k}\right)\right) \leq Q_2\left(\{\pi'_l\}_{l=1}^{k}\right). \tag{4.5}$$

Algorithm 4.2 Incremental k-means clustering algorithm (also see [DHS01], Section 10.8)

For a user-supplied tolerance $\mathtt{tol} < 0$ do the following.

1. Start with an arbitrary partitioning $\left\{\pi_l^{(0)}\right\}_{l=1}^{k}$. Set the index of iteration $t = 0$.

2. Generate the partition $\mathtt{nextFV}\left(\left\{\pi_l^{(t)}\right\}_{l=1}^{k}\right)$.

   If $\left[Q_2\left(\mathtt{nextFV}\left(\left\{\pi_l^{(t)}\right\}_{l=1}^{k}\right)\right) - Q_2\left(\left\{\pi_l^{(t)}\right\}_{l=1}^{k}\right) \leq \mathtt{tol}\right]$

   set $\left\{\pi_l^{(t+1)}\right\}_{l=1}^{k} = \mathtt{nextKM}\left(\left\{\pi_l^{(t)}\right\}_{l=1}^{k}\right)$
   increment t by 1.
   go to 2.

3. Stop.

EXAMPLE 4.2 *Let the vector set and the initial partition be given by Example 4.1. A single iteration of incremental k-means generates the optimal partition* $\pi_1^{(1)} = \{\mathbf{x}_1\}$, $\pi_2^{(1)} = \{\mathbf{x}_2, \mathbf{x}_3\}$ *as shown in the following figure.*

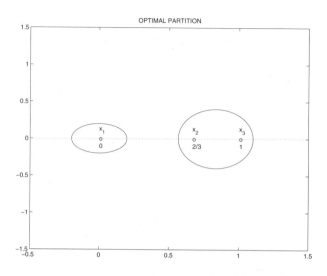

While computationally more accurate, incremental k-means is slower than batch k-means. Each iteration of incremental k-means changes cluster affiliation of a single vector only. The examples suggest the following "merger" of the two algorithms.

Unlike the means algorithm of [Kog01b] the algorithm described above keeps the number of clusters k fixed throughout the iterations. Otherwise the above algorithm enjoys advantages of the means algorithm:

1. The means algorithm always outperforms batch k-means in cluster quality (see [Kog01b]).

2. All numerical computations associated with Step 3 of the means algorithm have been already performed at Step 2 (see (4.3) and (4.4)). The improvement over batch k-means comes, therefore, at virtually no additional computational expense.

For simplicity we henceforth refer to Algorithm 4.3 as the means algorithm.

The k-means algorithm is known to be sensitive to the choice of an initial partition. A clustering algorithm that may be used for generating good initial partitions is presented next.

4.2.2   Principal Direction Divisive Partitioning

A memory-efficient and fast clustering algorithm was introduced recently by D. Boley [Bol98]. The method is not based on any distance or similarity measure, and takes advantage of sparsity of the "word by document" matrix.

The algorithm proceeds by dividing the entire collection into two clusters by using principal directions. Each of these two clusters will be divided into two subclusters using the same process recursively. The subdivision of a cluster is stopped when the cluster satisfies a certain "quality" criterion (e.g., the cluster's variance does not exceed a predefined threshold).

Algorithm 4.3 Simplified Version of the Means Clustering Algorithm (see [Kog01b]).

For user-supplied tolerances $\text{tol}_1 < 0$ and $\text{tol}_2 < 0$ do the following.

1. Start with an arbitrary partitioning $\left\{\pi_l^{(0)}\right\}_{l=1}^{k}$. Set the index of iteration $t = 0$.

2. Generate the partition $\text{nextKM}\left(\left\{\pi_l^{(t)}\right\}_{l=1}^{k}\right)$.

   If $\left[\mathcal{Q}_2\left(\text{nextKM}\left(\left\{\pi_l^{(t)}\right\}_{l=1}^{k}\right)\right) - \mathcal{Q}_2\left(\left\{\pi_l^{(t)}\right\}_{l=1}^{k}\right) \leq \text{tol}_1\right]$

   set $\left\{\pi_l^{(t+1)}\right\}_{l=1}^{k} = \text{nextKM}\left(\left\{\pi_l^{(t)}\right\}_{l=1}^{k}\right)$

   increment t by 1.

   go to 2.

3. Generate the partition $\text{nextFV}\left(\left\{\pi_l^{(t)}\right\}_{l=1}^{k}\right)$.

   If $\left[\mathcal{Q}_2\left(\text{nextFV}\left(\left\{\pi_l^{(t)}\right\}_{l=1}^{k}\right)\right) - \mathcal{Q}_2\left(\left\{\pi_l^{(t)}\right\}_{l=1}^{k}\right) \leq \text{tol}_2\right]$

   set $\left\{\pi_l^{(t+1)}\right\}_{l=1}^{k} = \text{nextFV}\left(\left\{\pi_l^{(t)}\right\}_{l=1}^{k}\right)$.

   increment t by 1.

   go to 2.

4. Stop.

Clustering of a set of vectors in $\mathbf{R}^n$ is, in general, a difficult task. There is, however, an exception. When $n = 1$, and all the vectors belong to a one-dimensional line, clustering becomes relatively easy. In many cases a good partition of a one-dimensional set Y into two subsets Y_1 and Y_2 amounts to a selection of a number, say μ, so that

$$Y_1 = \{y \: : \: y \in Y, \; y \leq \mu\}, \text{ and } Y_2 = \{y \: : \: y \in Y, \; y > \mu\} \qquad (4.6)$$

(in [Bol98], e.g., μ is the mean).

The basic idea of Boley's Principal Direction Divisive Partitioning (PDDP) algorithm is the following.

1. Given a set of vectors $\mathbf{X}$ in $\mathbf{R}^n$ determine the line Ł that approximates $\mathbf{X}$ in the "best possible way".

2. Project $\mathbf{X}$ onto Ł, and denote the projection of the set $\mathbf{X}$ by Y (note that Y is just a set of scalars). Denote the projection of a vector $\mathbf{x}$ by y.

3. Partition Y into two subsets Y_1 and Y_2 as described by Eq. (4.6).

4. Generate the induced partition $\{\mathbf{X}_1, \mathbf{X}_2\}$ of $\mathbf{X}$ as follows.

$$\mathbf{X}_1 = \{\mathbf{x} \ : \ y \in Y_1\}, \text{ and } \mathbf{X}_2 = \{\mathbf{x} \ : \ y \in Y_2\}. \qquad (4.7)$$

D. Boley has suggested the line that maximizes variance of the projections as the best one-dimensional approximation of an n-dimensional set. The direction of the line is defined by the eigenvector of the covariance matrix C corresponding to the largest eigenvalue. Since C is symmetric and positive semidefinite all the eigenvalues $\lambda_i, i = 1, 2, \ldots, n$ of the matrix are real and nonnegative; that is, $\lambda_1 \geq \lambda_2 \geq \cdots \geq \lambda_n \geq 0$. Furthermore, while the *scatter* value of the document set is $\lambda_1 + \lambda_2 + \cdots + \lambda_n$, the scatter value of the one-dimensional projection is only λ_1 (see [Bol98]). The quantity

$$\frac{\lambda_1}{\lambda_1 + \lambda_2 + \cdots + \lambda_n} \qquad (4.8)$$

may, therefore, be considered as the fraction of information preserved under the projection (in contrast with the "lost" information $(\lambda_2 + \cdots + \lambda_n)/(\lambda_1 + \lambda_2 + \cdots + \lambda_n)$). In spite of the fact that the numerator of (4.8) contains only one eigenvalue of a large matrix, the algorithm generates remarkable results (see, e.g., [Bol98, BGG+99a, BGG+99b]). For instance, examples provided in [Kog01b] show that an application of the k-means clustering algorithm to a partition generated by PDDP leads to only about 5% improvement in the objective function value.

In the next section, we describe the data set and corresponding feature selection problem considered in this study.

4.3   Data and Term Quality

Our data set is a merger of the three document collections (available from `http://www.cs.utk.edu/ lsi/`):

- DC0 (Medlars Collection 1033 medical abstracts),

- DC1 (CISI Collection 1460 information science abstracts),

- DC2 (Cranfield Collection (1398 aerodynamics abstracts).

The Cranfield collection tackled by Dhillon and Modha contained two empty documents. These two documents have been removed from DC2. The other document collections are identical.

We denote the overall collection of 3891 documents by DC. After stop-word removal (see `ftp://ftp.cs.cornell.edu/pub/smart/english.stop`), and stemming (see [Por80]) the data set contains 15,864 unique terms (no stemming was applied to the 24,574 unique words reported in [DM01]).

Our first goal is to select "good" index terms. We argue that for recovering the three document collections the term "blood" is much more useful than the term "case". Indeed, while the term "case" occurs in 253 Medlars documents, 72 CISI documents, and 365 Cranfield documents, the term "blood" occurs in 142 Medlars documents, 0 CISI documents, and 0 Cranfield documents. With each term $\mathbf{t}$ we associate a three-dimensional "direction" vector $\mathbf{d(t)} = (d_0(\mathbf{t}), d_1(\mathbf{t}), d_2(\mathbf{t}))$, so that $d_i(\mathbf{t})$ is the number of documents in a collection DCi containing the term $\mathbf{t}$. So, for example, $\mathbf{d}(\text{"case"}) = (253, 72, 365)$, and $\mathbf{d}(\text{"blood"}) = (142, 0, 0)$. In addition to "blood", terms "layer" ($\mathbf{d}(\text{"layer"}) = (6, 0, 358)$), or "retriev" ($\mathbf{d}(\text{"retriev"}) = (0, 262, 0)$) seem to be much more useful than the terms "case", "studi" and "found" with $\mathbf{d}(\text{"studi"}) = (356, 341, 238)$, and $\mathbf{d}(\text{"found"}) = (211, 93, 322)$, respectively.

When only the "combined" collection DC of 3891 documents is available the above-described construction of direction vectors is not possible. In Sections 4.4 and 4.5, we present algorithms that attempt to select "useful" terms when the direction vector $\mathbf{d(t)}$ is not available.

For each selection algorithm described in this chapter we introduce a quality functional q, so that the quality of a term $\mathbf{t}$ is given by $q(\mathbf{t})$. Higher values of $q(\mathbf{t})$ correspond to "better" terms $\mathbf{t}$. To exploit statistics of term occurrence throughout the corpus we remove terms that occur in less than r sentences across the collection, and denote the remaining terms by slice(r) (r should be collection dependent; the experiments in this chapter are performed with $r = 20$). The first l best quality terms that belong to slice(r) define the dimension of the vector space model.

In the next two sections, we present two different term-selection techniques along with corresponding document clustering results.

4.4   Term Variance Quality

We denote the frequency of a term $\mathbf{t}$ in the document $\mathbf{d}_j$ by f_j. Following the ideas of Salton and McGill [SM83] we measure the quality of the term $\mathbf{t}$ by

$$q_0(\mathbf{t}) = \sum_{j=1}^{n_0} f_j^2 - \frac{1}{n_0} \left[\sum_{j=1}^{n_0} f_j \right]^2, \tag{4.9}$$

where n_0 is the total number of documents in the collection (note that $q_0(\mathbf{t})$ is proportional to the term frequency variance). Tables 4.1 and 4.2 present 15 "best", and 15 "worst" terms for slice(20) in our collection of 3891 documents.

Term	$q_0(t)$	$d_0(t)$	$d_1(t)$	$d_2(t)$
flow	7687.795	35	34	714
librari	7107.937	0	523	0
pressur	5554.151	57	12	533
number	5476.418	92	204	568
cell	5023.158	210	2	2
inform	4358.370	28	614	44
bodi	3817.281	84	23	276
system	3741.070	82	494	84
wing	3409.713	1	0	216
effect	3280.777	244	159	539
method	3239.389	121	252	454
layer	3211.331	6	0	358
jet	3142.879	1	0	92
patient	3116.628	301	3	0
shock	3085.249	4	1	224

Table 4.1. Fifteen "Best" Terms in Slice(20) According to q_0

Term	$q_0(t)$	$d_0(t)$	$d_1(t)$	$d_2(t)$
suppos	21.875	6	7	9
nevertheless	21.875	6	11	5
retain	21.875	9	4	9
art	21.875	0	20	2
compos	21.875	5	5	12
ago	21.875	2	18	2
elabor	21.875	3	16	3
obviou	21.897	4	9	6
speak	20.886	6	12	3
add	20.886	3	14	4
understood	20.886	2	14	5
pronounc	20.886	18	0	3
pertain	19.897	3	8	9
merit	19.897	1	9	10
provis	19.897	1	18	1

Table 4.2. Fifteen "Worst" Terms in Slice(20) According to q_0

To evaluate the impact of feature selection by q_0 on clustering we conduct the following experiment. The best quality 600 terms are selected, and unit norm vectors for the 3891 documents are built (we use the tfn scheme to construct document vectors; for details see [DM01]). A two-step procedure is employed to partition the 3891 vectors into three clusters:

1. the PDDP algorithm is applied to generate 3 clusters (the obtained clusters are used as an initial partition in the next step); and

2. the means algorithm is applied to the partition obtained in the previous step.

Note that there is no a priori connection between document collection i and cluster i. Hence, one can not expect the confusion matrix to have diagonal structure unless rows (or columns) of the matrix are suitably permuted. A good clustering procedure should be able to produce a confusion matrix with a single "dominant" entry in each row. The confusion matrices for the three clusters provided in Tables 4.3 and 4.4 illustrate this remark.

	DC0	DC1	DC2
cluster 0	272	9	1379
cluster 1	4	1285	11
cluster 2	757	166	8
empty documents			
cluster 3	0	0	0

Table 4.3. PDDP-Generated Initial Confusion Matrix with 470 Misclassified Documents Using 600 Best q_0 Terms

When the number of terms is relatively small some documents may contain no selected terms, and their corresponding vectors would be zeros. We always remove these vectors ahead of clustering and assign the "empty" documents to a special cluster. This cluster concludes the confusion matrix (and is empty in this experiment).

	DC0	DC1	DC2
cluster 0	1	3	1365
cluster 1	8	1433	18
cluster 2	1024	24	15
empty documents			
cluster 3	0	0	0

Table 4.4. Means-Generated Final Confusion Matrix with 69 Misclassified Documents Using 600 Best q_0 Terms

While the quality of the confusion matrix presented above is similar to that reported in [DM01] (see Section 9.1), the dimension of our vector space model, 600, is about only 15% of the vector space dimension reported in [DM01].

The abstracts comprising the document collection DC are relatively short documents (from a half page to a page and a half long). It is not unusual to find terms that occur in many documents only once. Such terms score high by (4.9). At the same time these terms may lack any specificity. Indeed, the term "studi" with $\mathbf{d}(\text{"studi"}) = (356, 341, 238)$ is ranked 28th by q_0, and the term "present" with $\mathbf{d}(\text{"present"}) = (236, 314, 506)$ is ranked 35th. In order to penalize such terms, we modify (4.9) and introduce the quality of term $q_1(\mathbf{t})$ as the variance of $\mathbf{t}$ over documents that contain the term *at least once*. That is,

$$q_1(\mathbf{t}) = \sum_{j=1}^{n_1} f_j^2 - \frac{1}{n_1} \left[\sum_{j=1}^{n_1} f_j \right]^2, \tag{4.10}$$

where n_1 is the number of documents in which $\mathbf{t}$ occurs at least once, and $f_j \geq 1$, $j = 1, \ldots, n_1$. Tables 4.5 and 4.6 present the 15 "best", and the 15 "worst" q_1 terms for slice(20), respectively.

Term	$q_1(\mathbf{t})$	$d_0(\mathbf{t})$	$d_1(\mathbf{t})$	$d_2(\mathbf{t})$
librari	3147.074	0	523	0
flow	3146.048	35	34	714
number	2734.665	92	204	568
pressur	2528.225	57	12	533
cell	2225.177	210	2	2
inform	1851.231	28	614	44
bodi	1768.182	84	23	276
system	1518.877	82	494	84
shock	1490.113	4	1	224
jet	1463.569	1	0	92
theori	1341.363	23	117	452
method	1303.141	121	252	454
layer	1296.008	6	0	358
patient	1247.944	301	3	0
effect	1210.772	244	159	539

Table 4.5. Fifteen "Best" Terms in Slice(20) According to q_1

We select the best 600 terms and apply first the PDDP algorithm, and then the means algorithm to the corresponding 3891 vectors. The resulting confusion matrices are given in Tables 4.7 and 4.8. An increase in the number of selected terms does lead to a modest improvement in the quality of confusion matrices. In what follows, we summarize the improvement for term selections based on q_0 and q_1. Table 4.9 presents results for terms selected by q_0. The first row of Table 4.9 lists

Term	$q_1(t)$	$d_0(t)$	$d_1(t)$	$d_2(t)$
add	0.000	3	14	4
retain	0.000	9	4	9
reproduc	0.000	7	12	5
provis	0.000	1	18	1
pronounc	0.000	18	0	3
diminish	0.000	16	5	14
suppos	0.000	6	7	9
doubt	0.000	4	12	10
speak	0.000	6	12	3
context	0.000	7	45	1
understood	0.000	2	14	5
pertain	0.000	3	8	9
bring	0.000	8	30	8
ago	0.000	2	18	2
occasion	0.000	18	11	1

Table 4.6. Fifteen "Worst" Terms in Slice(20) According to q_1

	DC0	DC1	DC2
cluster 0	461	10	1380
cluster 1	3	803	0
cluster 2	569	647	18
"empty" documents			
cluster 3	0	0	0

Table 4.7. PDDP-Generated Initial Confusion Matrix with 1061 Misclassified Documents Using 600 Best q_1 Terms

	DC0	DC1	DC2
cluster 0	0	3	1360
cluster 1	6	1416	13
cluster 2	1027	41	25
empty documents			
cluster 3	0	0	0

Table 4.8. Means-Generated Final Confusion Matrix with 88 Misclassified Documents Using 600 Best q_1 Terms

clustering algorithms, and the first column shows the number of selected terms. The other columns indicate the number of misclassified documents. The displayed results indicate that the algorithm "collapses" when the number of selected terms drops below 600. Table 4.10 contains information relevant to q_1.

# of Terms	Documents Misclassified by	
	PDDP	Means
500	1062	989
600	470	69
700	388	63
1000	236	55
1300	181	53

Table 4.9. Number of Misclassified Documents for Term Selection Based on q_0

# of Terms	Documents Misclassified by	
	PDDP	Means
500	1055	94
600	1061	88
700	617	74
1000	410	64
1300	232	55

Table 4.10. Number of Misclassified Documents for Term Selection Based on q_1

The tables indicate that with 1300 selected terms (i.e., only about 30% of the 4099 terms reported in [DM01]) the number of "misclassified" documents is slightly lower than the number reported in [DM01].

In the next section, we introduce a measure of distance between terms. The distance is based on term co-occurrence in sentences across the document collection. The quality of a term **t** presented next is based on distribution of terms "similar" to **t** and co-occurring with **t** in sentences across the document collection.

4.5   Same Context Terms

The second approach to the term selection problem is based on the co-occurrence of "similar" terms in "the same context". Our departure point is the definition (attributed to Leibniz): two expressions are synonymous if the substitution of one for the other never changes the truth value of a sentence in which the substitution is made.

We follow the ideas of Grefenstette [Gre94]: "you can begin to know the meaning of a word (or term) by the company it keeps," and "words or terms that occur in 'the same context' are 'equivalent'," and Schütze and Pedersen [SP95]: "the assumption is that words with similar meanings will occur with similar neighbors if enough text material is available." Profiles introduced below formalize these notions.

4.5.1   Term Profiles

Our construction is the following.

1. Let $T = \{t_1, \ldots, t_m\}$ be an alphabetically sorted list of unique terms that occur in the document collection DC.

2. For each term t in T denote the set of sentences in DC containing t by $s(t)$. The size of the set is denoted by $|s(t)|$, and $s_{max} = \max_{t \in T} |s(t)|$.

3. For each term $t \in T$ the profile $\mathcal{P}(t)$ is defined next:

DEFINITION 4.3 *The profile* $\mathcal{P}(t)$ *of the term* t *is a set of terms from the list* T *that co–occur in sentences together with the term* t, *that is,*

$$\mathcal{P}(t) = \{t' : t' \in s(t)\}.$$

Profile $\mathcal{P}(t)$ contains corpus dependent information concerning the term t and the company it keeps. There are a number of ways to compute term similarity based on the respective profiles [Kog02]. A way to express the similarity is described below.

4. Let $\{s_1, \ldots, s_n\}$ be the set of all sentences contained in the document collection DC. The *term by sentence* matrix S is an $m \times n$ matrix whose entry S_{ij} is the number of times the term t_i occurs in the sentence s_j. The term t_i profile vector $P(t_i) = (P_1, \ldots, P_m)^T$ is the ith column of the symmetric matrix SS^T. The jth coordinate of the vector $P_j = \left(SS^T\right)_{ij}$ is the number of times the terms t_i and t_j co-occur in sentences across the document collection DC. Since $P_i \neq 0$, the vector $P(t_i)$ can be normalized.

5. DEFINITION 4.4 *Unit profile vector* $P(t)$ *of term* t *is defined to be* $(P(t))/\|P(t)\|$.

Words/terms with similar meanings (as per a given document collection) generate similar unit profile vectors (for details see [Kog02]). We next provide a formula for term quality based on term profile.

4.5.2   Term Profile Quality

The term profile quality $q_p(t)$ introduced in this section is based on the distribution of terms similar to t in the profile $\mathcal{P}(t)$.

For each $t' \in \mathcal{P}(t)$ compute the dot product $c' = P(t)^T P(t')$. We now sort the profile $\mathcal{P}(t)$ with respect to the dot products c', so that if $\mathcal{P}(t) = \{t_0, t_1, \ldots, t_n\}$, $(t_0 = t)$, then $1 = c_0 \geq c_1 \geq \cdots \geq c_n \geq 0$. We denote the frequency of the term

$\mathbf{t}_i$ in the profile $\mathcal{P}(\mathbf{t})$ by f_i and define the term profile quality $q_\mathbf{p}(\mathbf{t}_0)$ by a somewhat contrived formula (justification is given below):

$$q_\mathbf{p}(\mathbf{t}_0) = \left[\frac{|\mathbf{s}(\mathbf{t}_0)|}{s_{max}} \right]^{0.2} \frac{1}{f_0 + ke(f)} \sum_{i=1}^{k} \left[f_i c_i \times \left(1 - \sqrt{\frac{|e(f) - f_i|}{ke(f)}} \right) \right], \quad (4.11)$$

where $e(f) = (1/k) \sum_{i=1}^{k} f_i$, and in this experiment $k = 2$. We note the following concerning the expression for the profile quality $q_\mathbf{p}$.

1. Due to the small power 0.2, the term $[(|\mathbf{s}(\mathbf{t}_0)|)/s_{max}]^{0.2}$ penalizes very frequent collection terms;

2. the normalizing term $1/(f_0 + ke(f))$ attempts to suppress the importance of very frequent profile terms similar to $\mathbf{t}_0$;

3. the term $f_i c_i$ reflects the measure of similarity between $\mathbf{t}_0$ and $\mathbf{t}_i$; and

4. the term $1 - \sqrt{(|e(f) - f_i|)/ke(f)}$ imposes a penalty on a term's deviation from the expected frequency.

Term	$q_\mathbf{p}(\mathbf{t})$	$d_0(\mathbf{t})$	$d_1(\mathbf{t})$	$d_2(\mathbf{t})$
laminar	0.264	0	0	231
layer	0.205	6	0	358
number	0.204	92	204	568
septal	0.202	25	0	0
free-stream	0.195	0	0	97
boundari	0.195	0	7	413
nephrectomi	0.168	23	0	0
unilater	0.161	27	0	0
defect	0.157	64	5	4
reynold	0.152	0	1	197
mach	0.141	0	0	384
nomin	0.136	2	2	17
moment	0.135	4	4	89
autom	0.128	0	46	0
biliari	0.126	17	0	0

Table 4.11. Fifteen "Best" Terms in Slice(20) According to $q_\mathbf{p}$

Tables 4.11 and 4.12 present 15 "best", and 15 "worst" terms for slice(20). For clustering purposes we selected 1000 best quality index terms. Although each selected term is contained in at least 20 sentences, the selected 1000 unit profile vectors $\mathbf{P}(\mathbf{t})$ of dimension 15,864 (which is the total number of the unique terms; see Section 4.3) are sparse. The average number of nonzero entries in a unit profile vector is 617, that is, less than 4% of 15,864, the dimension of the vector space.

Term	$q_{\mathbf{p}}(\mathbf{t})$	$d_0(\mathbf{t})$	$d_1(\mathbf{t})$	$d_2(\mathbf{t})$
determin	0.001	108	116	299
larg	0.001	80	175	201
approxim	0.001	31	46	377
found	0.001	211	93	322
analysi	0.001	42	184	276
includ	0.001	75	169	225
rate	0.001	111	49	145
paper	0.001	31	265	200
experi	0.001	105	133	152
result	0.000	278	288	692
effect	0.000	244	159	539
studi	0.000	356	341	238
gener	0.000	76	311	329
develop	0.000	176	366	264
case	0.000	253	72	365

Table 4.12. Fifteen "Worst" Terms in Slice(20) According to $q_{\mathbf{p}}$

Next we apply the means algorithm to partition 1000 term vectors into three term clusters $\mathbf{T}_0$, $\mathbf{T}_1$, and $\mathbf{T}_2$. The partition of the document collection DC is based on the term clusters. For each document $\mathbf{d}$ we construct a three-dimensional vector $(t_0(\mathbf{d}), t_1(\mathbf{d}), t_2(\mathbf{d}))$, where $t_i(\mathbf{d})$ is the number of terms from term cluster $\mathbf{T}_i$ contained in the document. The document $\mathbf{d}$ belongs to document cluster i if

$$t_i(\mathbf{d}) \geq t_j(\mathbf{d}), \quad j = 0, 1, 2.$$

The confusion matrix for this partition of 3891 documents with 105 misclassified documents is given in Table 4.13. A 30% increase in the number of index terms leads to the decrease in the number of misclassified documents to 94, and eliminates empty documents.

	DC0	DC1	DC2
cluster 0	23	32	1386
cluster 1	975	3	1
cluster 2	35	1424	11
empty documents			
cluster 3	0	1	0

Table 4.13. Means Generated Final Confusion Matrix with 105 Misclassified Documents

The clustering algorithms discussed so far deal with general vector sets in $\mathbf{R}^n$. In the next section we present a clustering algorithm specifically designed to handle unit norm document vectors.

4.6   Spherical Principal Directions Divisive Partitioning

In this section we mimic the simple and elegant idea due to Boley and approximate a set of unit vectors $\mathbf{X} \subset \mathbf{R}^n$ by a one-dimensional great circle of $\mathbf{S}^{n-1}$. A great circle is represented by an intersection of $\mathbf{S}^{n-1}$ and a two dimensional subspace $\mathbf{P}$ of $\mathbf{R}^n$. The proposed algorithm is the following.

Algorithm 4.4 Spherical Principal Directions Divisive Partitioning (sPDDP) Clustering Algorithm

---

1. Given a set of unit vectors $\mathbf{X}$ in $\mathbf{R}^n$ determine the two dimensional plane $\mathbf{P}$ that approximates $\mathbf{X}$ in the "best possible way."

2. Project $\mathbf{X}$ onto $\mathbf{P}$. Denote the projection of the set $\mathbf{X}$ by $\mathbf{Y}$, and the projection of a vector $\mathbf{x}$ by $\mathbf{y}$ (note that $\mathbf{y}$ is two-dimensional).

3. If $\mathbf{y} \neq 0$ "push" $\mathbf{y} \in \mathbf{Y}$ to the great circle, and denote the corresponding vector by $\mathbf{z} = \mathbf{y}/\|\mathbf{y}\|$. Denote the constructed set by $\mathbf{Z}$.

4. Partition $\mathbf{Z}$ into two clusters $\mathbf{Z}_1$ and $\mathbf{Z}_2$. Assign projections $\mathbf{y}$ with $\|\mathbf{y}\| = 0$ to $\mathbf{Z}_1$.

5. Generate the induced partition $\{\mathbf{X}_1, \mathbf{X}_2\}$ of $\mathbf{X}$ as follows.

$$\mathbf{X}_1 = \{\mathbf{x} \ : \ \mathbf{z} \in \mathbf{Z}_1\}, \text{ and } \mathbf{X}_2 = \{\mathbf{x} \ : \ \mathbf{z} \in \mathbf{Z}_2\}. \qquad (4.12)$$

If, following ideas of [Bol98], the best two-dimensional approximation of the document set is the plane $\mathbf{P}$ that maximizes variance of the projections, then $\mathbf{P}$ is defined by two eigenvectors of the covariance matrix C corresponding to the largest eigenvalues λ_1 and λ_2. The "preserved" information under this projection is

$$\frac{\lambda_1 + \lambda_2}{\lambda_1 + \lambda_2 + \cdots + \lambda_n}. \qquad (4.13)$$

Note that the quantity given by (4.13) may be almost twice as much as the preserved information under the projection on the one-dimensional line given by (4.8). As we show later in this section this may lead to a significant improvement over results provided by PDDP.

4.6.1   Two-Cluster Partition of Vectors on the Unit Circle

We now describe in detail Step 4 of the sPDDP algorithm. Specifically we are concerned with the following problem. Given a set of unit vectors $\mathbf{Z} = \{\mathbf{z}_1, \ldots, \mathbf{z}_m\} \subset \mathbf{R}^2$ partition $\mathbf{Z}$ into two "optimal" clusters π_1^o and π_2^o.

A straightforward imitation of Boley's construction leads to the following solution. If $\mathbf{z} = \mathbf{z}_1 + \cdots + \mathbf{z}_m \neq 0$, then the line defined by $\mathbf{z}$ cuts the plane into two half-planes. The subset of $\mathbf{Z}$ that belongs to the left half-plane is denoted by $\mathbf{Z}_-$, and the subset of $\mathbf{Z}$ that belongs to the right half-plane is denoted by $\mathbf{Z}_+$. If $\mathbf{z}$ is zero, then, in order to generate the partition, we choose an arbitrary line passing through the origin.

Lack of robustness is, probably, the most prominent drawback of the suggested partitioning. Indeed, let $\{\mathbf{z}_1, \ldots, \mathbf{z}_m\}$ be a set of unit vectors concentrated around, say, the vector $\mathbf{e}_1 = (1, 0)^T$. If the set $\mathbf{Z}$ contains vectors $\{\mathbf{z}_1, \ldots, \mathbf{z}_m\}$ and their negatives $\{-\mathbf{z}_1, \ldots, -\mathbf{z}_m\}$, then $\mathbf{z} = 0$. This $\mathbf{z}$ does not do much to recover "good" clusters (although $\pi_1 = \{\mathbf{z}_1, \ldots, \mathbf{z}_m\}$, and $\pi_2 = \{-\mathbf{z}_1, \ldots, -\mathbf{z}_m\}$ look like a reasonable partition; see figure below).

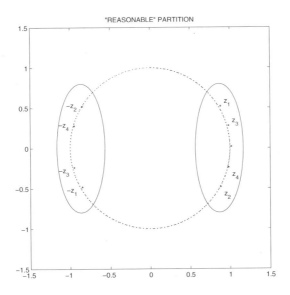

Things get worse when $\mathbf{e}_1$ is assigned to the vector set $\mathbf{Z}$; that is, $\mathbf{Z} = \{\mathbf{z}_1, \ldots, \mathbf{z}_m, -\mathbf{z}_1, \ldots, -\mathbf{z}_m, \mathbf{e}_1\}$. Now $\mathbf{z} = \mathbf{e}_1$, and regardless of how "densely" the vectors $\{\mathbf{z}_1, \ldots, \mathbf{z}_m\}$ are concentrated around $\mathbf{e}_1$ the clusters $\mathbf{Z}_+$ and $\mathbf{Z}_-$ most probably contain vectors from both sets $\{\mathbf{z}_1, \ldots, \mathbf{z}_m\}$ and $\{-\mathbf{z}_1, \ldots, -\mathbf{z}_m\}$. This poor partition is illustrated by the figure below.

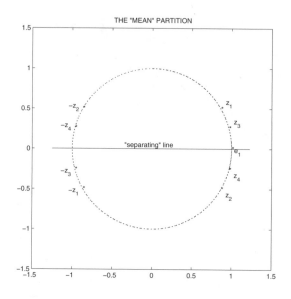

To define an optimal partition we measure the quality of a partition $\{\pi_1, \pi_2\}$ by the "spherical" objective function

$$\mathcal{Q}_s\left(\{\pi_1, \pi_2\}\right) = \left\|\sum_{\mathbf{z} \in \pi_1} \mathbf{z}\right\| + \left\|\sum_{\mathbf{z} \in \pi_2} \mathbf{z}\right\| \tag{4.14}$$

introduced by Dhillon and Modha [DM01]. Denote an optimal partition, that is, one that maximizes (4.14), by $\{\pi_1^o, \pi_2^o\}$. It can be seen that for each optimal partition $\{\pi_1^o, \pi_2^o\}$ there is a nonzero vector $\mathbf{x}^o$ so that the clusters π_1^o and π_2^o are separated by the line passing through the origin and defined by $\mathbf{x}^o$ (see [DM01]).

Since each unit vector $\mathbf{z} \in \mathbf{R}^2$ can be uniquely represented by $e^{i\theta}$ with $0 \le \theta < 2\pi$ the associated clustering problem is essentially one-dimensional. We denote $\mathbf{z}_j$ by $e^{i\theta_j}$, and assume (without any loss of generality), that

$$0 \le \theta_1 \le \theta_2 \le \cdots \le \theta_m < 2\pi.$$

As in the case of clustering points on a line, it is tempting to assume that for some j a line passing through the origin and midway between $\mathbf{z}_j$ and $\mathbf{z}_{j+1}$ recovers the optimal partition. We show by the following example that this is not necessarily the case.

EXAMPLE 4.3 *Let* $\mathbf{z}_1 = (1, 0)^T$, $\mathbf{z}_2 = (\cos(2\pi/3 - \epsilon), \sin(2\pi/3 - \epsilon))^T$, $\mathbf{z}_3 = -\mathbf{z}_1$, *and* $\mathbf{z}_4 = (\cos(-2\pi/3 + \epsilon), \sin(-2\pi/3 + \epsilon))^T$. *It is easy to see that when* $\epsilon = 0$ *the optimal partition is* $\{\pi_1^o, \pi_k^o\} = \{\{\mathbf{z}_1\}, \{\mathbf{z}_2, \mathbf{z}_3, \mathbf{z}_4\}\}$ *with* $\mathcal{Q}_s\left(\{\pi_1^o, \pi_2^o\}\right) = 3$.

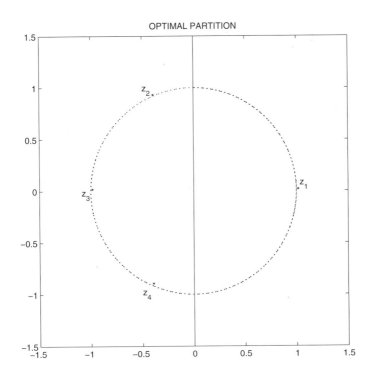

While a small positive ϵ (e.g., $\epsilon = \pi/36$) does not change the optimal partition, the four midpoint lines generate clusters containing two vectors each (a partition i is generated by a line passing through the origin and the midpoint between $\mathbf{z}_i$ and $\mathbf{z}_{i+1}$). These partitions do not contain the optimal partition $\{\pi_1^o, \pi_2^o\}$. We next present the four midpoint line partitions with $\epsilon = \pi/36$.

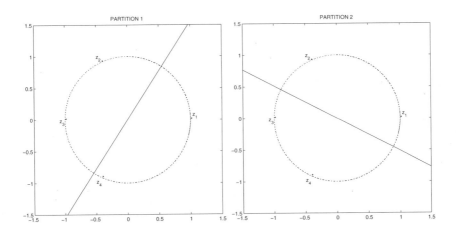

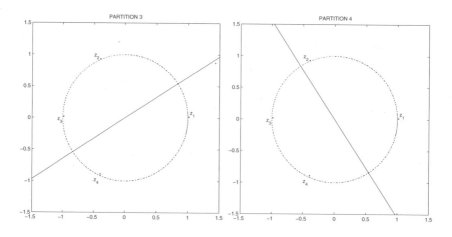

To analyze the failure of Example 4.3, and to propose a remedy we introduce the formal definition of the left and right half-planes generated by a vector $\mathbf{x}$, and describe a procedure that computes the optimal separator $\mathbf{x}^o$.

- For a nonzero vector $\mathbf{x} \in \mathbf{R}^2$ we denote by $\mathbf{x}^\perp$ the vector obtained from $\mathbf{x}$ by rotating it clockwise by an angle of 90^0; that is,

$$\mathbf{x}^\perp = \begin{bmatrix} 0 & 1 \\ -1 & 0 \end{bmatrix} \mathbf{x}.$$

- For a nonzero vector $\mathbf{x} \in \mathbf{R}^2$, and a set of vectors $\mathbf{Z} = \{\mathbf{z}_1, \ldots, \mathbf{z}_m\} \subset \mathbf{R}^2$ define two subsets of $\mathbf{Z}$ — the positive $\mathbf{Z}_+(\mathbf{x}) = \mathbf{Z}_+$, and the negative $\mathbf{Z}_-(\mathbf{x}) = \mathbf{Z}_-$ as follows.

$$\mathbf{Z}_+ = \{\mathbf{z} : \mathbf{z} \in \mathbf{Z},\ \mathbf{z}^T \mathbf{x}^\perp \geq 0\} \text{ and } \mathbf{Z}_- = \{\mathbf{z} : \mathbf{z} \in \mathbf{Z},\ \mathbf{z}^T \mathbf{x}^\perp < 0\}. \quad (4.15)$$

- For two unit vectors $\mathbf{z}' = e^{i\theta'}$ and $\mathbf{z}'' = e^{i\theta''}$ we denote the "midway" vector $e^{i(\theta'+\theta'')/2}$ by $\mathtt{mid}\,(\mathbf{z}', \mathbf{z}'')$.

As the optimal separating line in Example 4.3 is rotated clockwise to $\mathtt{mid}\,(\mathbf{z}_2, \mathbf{z}_1)$ it crosses $\mathbf{z}_4$ changing cluster affiliation of this vector (see figures below).

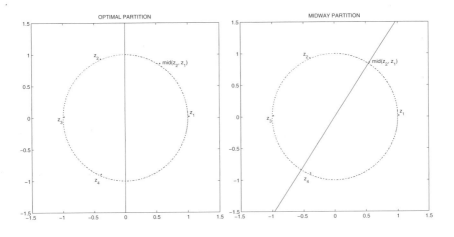

This could have been prevented if instead of rotating the optimal separating line all the way to $\mathtt{mid}\,(\mathbf{z}_2, \mathbf{z}_1)$ one had rotated it to $\mathtt{mid}\,(\mathbf{z}_2, -\mathbf{z}_4)$. The optimal separating line and the line passing through $\mathtt{mid}\,(\mathbf{z}_2, -\mathbf{z}_4)$ and the origin generate identical partitions (see Figure 4.1).

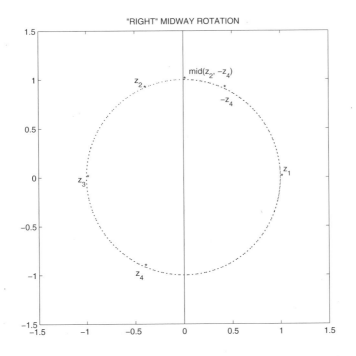

Figure 4.1. Optimal partition.

In general, if the set $\mathbf{Z} = \{\mathbf{z}_1 = e^{i\theta_1}, \ldots, \mathbf{z}_m = e^{i\theta_m}\}$ is symmetric with respect to the origin, (i.e., for each $\mathbf{z}_i \in \mathbf{Z}$ there exists $\mathbf{z}_l \in \mathbf{Z}$ such that $\mathbf{z}_i = -\mathbf{z}_l$), then for

$$\mathbf{x}' = e^{i\theta'}, \ \mathbf{x}'' = e^{i\theta''}, \ \text{with } \theta_j < \theta' \le \theta'' < \theta_{j+1},$$

the partitions

$$\{\mathbf{Z}_+(\mathbf{x}'), \mathbf{Z}_-(\mathbf{x}')\}, \ \text{and } \{\mathbf{Z}_+(\mathbf{x}''), \mathbf{Z}_-(\mathbf{x}'')\}$$

are identical. This observation suggests the following simple procedure for recovering the optimal partition $\{\pi_1^o, \pi_2^o\}$.

1. Let $\mathbf{W} = \{\mathbf{w}_1, \ldots, \mathbf{w}_m, \mathbf{w}_{m+1}, \ldots, \mathbf{w}_{2m}\}$ be a set of two-dimensional vectors defined as follows.

$$\mathbf{w}_i = \mathbf{z}_i \text{ for } i = 1, \ldots, m, \text{ and } \mathbf{w}_i = -\mathbf{z}_i \text{ for } i = m+1, \ldots, 2m.$$

2. If needed, reassign indices so that

$$\mathbf{w}_j = e^{i\theta_j}, \text{ and } 0 \le \theta_1 \le \theta_2 \le \cdots \le \theta_{2m} < 2\pi.$$

3. With each subscript j associate a partition $\{\pi_1^j, \pi_2^j\}$ of $\mathbf{Z}$ as follows:

   (a) set $\mathbf{x} = (\mathbf{w}_j + \mathbf{w}_{j+1})/2$;
   (b) set $\pi_1^j = \mathbf{Z}_+(\mathbf{x})$, and $\pi_2^j = \mathbf{Z}_-(\mathbf{x})$.

   Note that:

   (a) The indices j and $j + m$ generate identical partitions. We, therefore, have to consider at most m distinct partitions generated by $j = 1, \ldots, m$.
   (b) The optimal partition that maximizes (4.14) is among the generated ones.

4. With each partition $\{\pi_1^j, \pi_2^j\}$ associate the value of the objective function $\mathcal{Q}_s^j = \mathcal{Q}_s\left(\{\pi_1^j, \pi_2^j\}\right)$. Let $\mathcal{Q}_s^k = \max\limits_{j=1,\ldots,m} \mathcal{Q}_s^j$; then the desired partition of $\mathbf{Z}$ is $\{\pi_1^o, \pi_2^o\} = \{\pi_1^k, \pi_2^k\}$.

4.6.2   Clustering with sPDDP

In what follows, we display clustering results for the document collection DC described in Section 4.3. To compare the results with those presented in Section 4.4, we select the 600 best q_0 quality terms (see Eq. (4.9)) to build document vectors. The confusion matrix for the three-cluster partition generated by sPDDP is given in Table 4.14 below. We remark that the confusion matrix is a significant improvement over the result presented in Table 4.3. A subsequent application of the means algorithm to the partition generated by sPDDP leads to a minor improvement of the result both in terms of confusion matrices, as well as in terms of the objective function $\mathcal{Q}_2$ (see Table 4.15).

	DC0	DC1	DC2
cluster 0	1000	3	1
cluster 1	8	10	1376
cluster 2	25	1447	21
empty documents			
cluster 3	0	0	0

Table 4.14. sPDDP-Generated Initial Confusion Matrix with 68 Misclassified Documents; Partition Quality $Q_2 = 3630.97$

	DC0	DC1	DC2
cluster 0	1023	21	10
cluster 1	1	3	1370
cluster 2	9	1436	18
empty documents			
cluster 3	0	0	0

Table 4.15. Means-Generated Final Confusion Matrix with 62 Misclassified documents; Partition Quality $Q_2 = 3630.38$

Table 4.16 summarizes clustering results for the sPDDP algorithm combined with the means-clustering algorithm for different choices of index terms (all term selections are based on the q_0 criterion). Note that while the combination of the

	Documents Misclassified by	
# of Terms	PDDP	Means
300	228	100
400	88	80
500	76	62
600	68	62

Table 4.16. Number of Misclassified Documents for Term Selection Based on q_0

PDDP and the means algorithms collapses when the number of selected terms drops below 600 (see Table 4.9), the combination of the sPDDP and the means algorithms performs reasonably well even when the number of selected terms is only 300.

Clustering results for different choices of index terms based on the q_1 criterion are similar to those presented above. The results are summarized in Table 4.17.

	Documents Misclassified by	
# of Terms	PDDP	Means
300	224	101
400	91	86
500	74	71
600	71	68

Table 4.17. Number of Misclassified Documents for Term Selection Based on q_1

4.7   Future Research

This chapter presents preliminary results concerning two information retrieval-related problems:

1. feature selection, and

2. document clustering.

We plan to further investigate profile-based term selection techniques as well as techniques based on term distribution across documents [GK02], and to run term selection experiments on larger document collections.

Clustering experiments with seven different objective functions reported by Zhao and Karypis [ZK02] indicate that the objective function based on cosine similarity (and used in [DM01]) "leads to the best solutions irrespective of the number of clusters for most of the data sets." We intend to combine the Spherical Principal Direction Divisive Partitioning algorithm with the modification of the spherical k-means algorithm recently reported by [DGK02].

The Spherical Principal Directions Divisive Partitioning algorithm introduced in the chapter utilizes the unit norm constraint imposed on document vectors. In many data mining applications, vectors representing data are normalized. For example:

1. in bioinformatics applications, fingerprint data are transformed to have mean zero and variance one, a fixed l_2 norm, or a fixed l_∞ norm [SS02];

2. in contemporary k-means type frameworks for word clustering, a word is represented by a discrete probability distribution, that is, by a vector of l_1 unit norm [DMK02, BB02, ST01];

3. the n-gram technique leads to a vector space model where document vectors have l_1 unit norm [Dam95]. The technique is proved to be useful in information retrieval applications [PN96], as well as in bioinformatics [GKSR+02].

We plan to derive and investigate clustering algorithms utilizing special constraints (among them l_p constraints mentioned above) imposed upon vector data sets.

While this chapter deals with a vector space model based on word occurrence across documents, additional research directions include clustering of vectors whose components are the frequencies of their distinct constituent n-grams

[Dam95]. The *n*-gram representation of a document is sparse, simple, and language independent. The sparsity of the vectors lends itself to processing with numerical linear algebra tools, although the matrices themselves may be much larger. We believe that the best clustering results may be achieved by combining a number of different techniques.

Acknowledgments: The authors thank Robert Libier for valuable suggestions that improved exposition of the results. The work of Dhillon was supported by NSF grant No. ACI-0093404, and grant no. 003658-0431-2001 from the Texas Higher Education Coordinating Board. The work of Kogan and Nicholas was supported in part by the US Department of Defense.

References

[BB99]   M.W. Berry and M. Browne.*Understanding Search Engines: Mathematical Modeling and Text Retrieval.*SIAM, Philadelphia, 1999.

[BB02]   P. Berkhin and J.D. Becher.Learning simple relations: Theory and applications.In *Proceedings of the Second SIAM International Conference on Data Mining*, Arlington, VA, pages 410-436, April 2002.

[BGG$^+$99a]   D. Boley, M. Gini, R. Gross, E.-H. Han, K. Hastings, G. Karypis, V. Kumar, B. Mobasher, and J. Moore.Document categorization and query generation on the World Wide Web using WebACE.*AI Review*, 13(5,6):365–391, 1999.

[BGG$^+$99b]   D. Boley, M. Gini, R. Gross, E.-H. Han, K. Hastings, G. Karypis, V. Kumar, B. Mobasher, and J. Moore.Partitioning-based clustering for Web document categorization.*Decision Support Systems*, 27(3):329–341, 1999.

[Bol98]   D.L. Boley.Principal direction divisive partitioning.*Data Mining and Knowledge Discovery*, 2(4):325–344, 1998.

[Dam95]   M. Damashek.Gauging similarity with n-grams: Language-independent categorization of text.*Science*, 267:843–848, 1995.

[DGK02]   I.S. Dhillon, Y. Guan, and J. Kogan.Refining clusters in high-dimensional text data.In *Proceedings of the Workshop on Clustering High Dimensional Data and Its Applications at the Second SIAM International Conference on Data Mining*, I.S. Dhillon and J. Kogan, eds., pages 71–82. SIAM, Philadelphia, 2002.

[DM01]   I.S. Dhillon and D.S. Modha.Concept decompositions for large sparse text data using clustering.*Machine Learning*, 42(1):143–175, Jan 2001.Also appears as IBM Research Report RJ 10147, Jul 1999.

[DMK02]   I.S. Dhillon, S. Malella, and R. Kumar.Enhanced word clustering for hierarchical text classification.In *KDD-2002*, 2002.

[DHS01]   R.O. Duda, P.E. Hart, and D.G. Stork.*Pattern Classification*, second edition.Wiley, New York, 2001.

[For65]   E. Forgy.Cluster analysis of multivariate data: Efficiency vs. interpretability' of classifications.*Biometrics*, 21(3):768, 1965.

[GK02]   E. Gendler and J. Kogan.Index terms selection for clustering large text data.In *Proceedings of the Workshop on Text Mining at the Second SIAM International Conference on Data Mining*, M.W. Berry, ed., pages 87–94, 2002.

[GKSR⁺02] M. Ganapathiraju, J. Klein-Seetharaman, R. Rosenfeld, J. Carbonell, and R. Reddy.Rare and frequent n-grams in whole-genome protein sequences.In *Proceedings of RECOMB'02: The Sixth Annual International Conference on Research in Computational Molecular Biology*, 2002.

[Gre94]   G. Grefenstette.*Explorations in Automatic Thesaurus Discovery*.Kluwer Academic, Boston, 1994.

[Kog01a]   J. Kogan.Clustering large unstructured document sets.In *Computational Information Retrieval*, M.W. Berry, ed., pages 107–117, SIAM, Philadelphia, 2001.

[Kog01b]   J. Kogan.Means clustering for text data.In *Proceedings of the Workshop on Text Mining at the First SIAM International Conference on Data Mining*, M.W. Berry, ed., pages 47–57, 2001.

[Kog02]   J. Kogan.Computational information retrieval.*Springer-Verlag Lecture Notes in Contributions to Statistics*, H.R. Lerche, ed., 2002. To appear.

[PN96]   C. Pearce and C. Nicholas.TELLTALE: Experiments in a dynamic hypertext environment for degraded and multilingual data.*Journal of the American Society for Information Science*, 47:263–275, 1996.

[Por80]   M.F. Porter.An algorithm for suffix stripping.*Program*, 14:130–137, 1980.

[SM83]   G. Salton and M.J. McGill.*Introduction to Modern Information Retrieval*.Mc Graw-Hill, New York, 1983.

[SP95]   H. Schütze and J. Pedersen.Information retrieval based on word senses.In *Proceedings of the Fourth Annual Symposium on Document Analysis and Information Retrieval*, Las Vegas, pages 161–175, 1995.

[SS02]   R. Shamir and R. Sharan.Algorithmic approaches to clustering gene expression data.In *Current Topics in Computational Molecular Biology*, T. Jiang, T. Smith, Y. Xu, and M. Q. Zhang, eds., pages 269–300, MIT Press, Cambridge, MA, 2002.

[ST01]   N. Slonim and N. Tishby.The power of word clusters for text classification.In *Proceedings of the 23rd European Colloquium on Information Retrieval Research (ECIR)*, Darmstadt, 2001.

[ZK02]   Y. Zhao and G. Karypis.Comparison of agglomerative and partitional document clustering algorithms.In *Proceedings of the Workshop on Clustering High Dimensional Data and Its Applications at the Second SIAM International Conference on Data Mining*, I.S. Dhillon and J. Kogan, eds., pages 83–93. SIAM, Philadelphia, 2002.

Part II

Information Extraction and Retrieval

5

Vector Space Models for Search and Cluster Mining

Mei Kobayashi
Masaki Aono

Overview

This chapter consists of two parts: a review of search and cluster mining algorithms based on vector space modeling followed by a description of a prototype search and cluster mining system. In the review, we consider *Latent Semantic Indexing* (LSI), a method based on the Singular Value Decomposition (SVD) of the document attribute matrix and *Principal Component Analysis* (PCA) of the document vector covariance matrix. In the second part, we present novel techniques for mining major and minor clusters from massive databases based on enhancements of LSI and PCA and automatic labeling of clusters based on their document contents. Most mining systems have been designed to find major clusters and they often fail to report information on smaller minor clusters. Minor cluster identification is important in many business applications, such as detection of credit card fraud, profile analysis, and scientific data analysis. Another novel feature of our method is the recognition and preservation of naturally occurring overlaps among clusters. Cluster overlap analysis is important for multiperspective analysis of databases. Results from implementation studies with a prototype system using over 100,000 news articles demonstrate the effectiveness of search and clustering engines.

5.1   Introduction

Public and private institutions are being overwhelmed with processing information in massive databases having documents of heterogeneous formats [WF99]. Since the documents are generated by different people or by machines, they are of heterogeneous format and may contain different types of data (text, audio, image, video, HTML) and text in different languages. Some successful methods for retrieving nuggets of information have been developed by researchers in the data mining community.[1] In this chapter, we focus on vector space modeling (VSM), an effective tool for information retrieval (IR) introduced by Salton and his colleagues [Sal71] over three decades ago. We examine means for enhancing the scalability of the method to enable mining information from massive databases orders of magnitude greater than originally envisioned by Salton. Several properties make VSM an attractive method. For instance:

- it can handle documents of heterogeneous format;
- it can handle different types of of multimedia data;
- it can process documents in many different languages;
- the IR process can be fully automated; and
- most of the computational workload can be carried out during the preprocessing stage so that query processing can take place in real-time.

In the second part of this chapter, we introduce a novel prototype system we developed that uses our new technique based on VSM for mining and labeling clusters in massive databases. Our technique is novel in that it can find and automatically label both *major* and *minor* clusters.[2] Although successful techniques have been developed to identify major clusters and their main themes in these massive databases, few have been developed for understanding smaller *minor* clusters. Furthermore, topics in major clusters of very large databases are often known even before mining technologies are used, either from direct experience with customers or from observation of market trends. In contrast, topics in minor clusters are not known since they are more difficult to spot from daily experience. Recently, corporate, government, and military planners are recognizing that mining even a portion of the information in minor clusters can be extremely valuable [SY02]. For example: corporations may want to mine customer data to find minor reasons for customer dissatisfaction (in addition to major reasons), since minor clusters may represent emerging trends or long-term small dissatisfactions that lead users to switch to another product; credit card and insurance firms may want to better understand customer data to set interest and insurance rates; security agencies may want to use mining technologies for profile analysis; and scientists may want to mine weather and geographical data to refine their forecasts and predictions of natural disasters.

The remainder of this chapter is organized as follows. In the next section, we review basic terminology and mathematical tools used in VSM, then introduce some methods for increasing the scalability of IR systems based on VSM. In Section 5.3 we examine clustering, another approach for addressing the scalability problem associated with VSMs. We review nonpartitioning approaches to mine clusters, and we propose two new algorithms for mining major and minor clusters. Results

---

[1] http://www.kdnuggets.com.

[2] *Major* and *minor* clusters are relatively *large* and relatively *small* clusters with respect to a database under consideration.

from implementation studies of our algorithms are presented in the penultimate section. This chapter concludes with a summary of findings and a discussion of possible directions for future research.

5.2   Vector Space Modeling (VSM)

5.2.1   The Basic VSM Model for IR

VSM has become a standard tool in IR systems since its introduction over three decades ago [BDJ99, BDO95, Sal71]. One of the advantages of the method is that it enables relevance ranking of documents of heterogeneous format (e.g., text, multilingual text, image audio, video) with respect to user input queries as long as the attributes are well-defined characteristics of the documents.

| 1 | 0 | 0 | 1 | 0 | 1 | 1 | 1 | 0 | . | . | . | 0 | 1 |

Document Vector - Boolean Model

| .1 | 0 | 0 | .2 | 0 | .3 | .8 | 1 | 0 | . | . | . | 0 | .11 |

Document Vector - Term-Weighted Model

Figure 5.1. Example of a document vector in binary format (top) and term-weighted format (bottom).

In *Boolean* vector models each coordinate of a document vector is zero (when the corresponding attribute is absent) or unity (when the corresponding attribute is present). *Term weighting* is a widely used refinement of Boolean models that takes into account the frequency of appearance of attributes (such as keywords and key phrases), the total frequency of appearance of each attribute in the document set, and the location of appearance (e.g., in the title, section header, abstract, or text). In our studies, we use a fairly common type of *Term Frequency Inverse Document Frequency weighting (TF-IDF)*, in which the weight of the ith term in the jth document, denoted by weight(i, j), is defined by

$$\text{weight}(i, j) = \begin{cases} (1 + tf_{i,j}) \log_2(n/df_i), & \text{if } tf_{i,j} \geq 1, \\ 0, & \text{if } tf_{i,j} = 0, \end{cases}$$

where $tf_{i,j}$ is defined as the number of occurrences of the ith term within the jth document d_j, and df_i is the number of documents (out of n) in which the term appears [MS00]. There are many variations of the *TF-IDF* formula. All are based on the idea that the term-weighting should reflect the relative importance of a term in a document (with respect to other terms in the document) as well as how important the term is in other documents. *TF-IDF* models assume that *importance* of a term in a document is reflected by its *frequency of appearance* in documents.[3] An example of a document vector in binary and term weighted format is given in Figure 5.1.

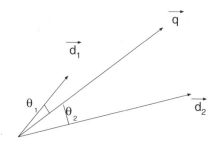

Figure 5.2. Similarity ranking of two documents d_1 and d_2 with respect to query q. Note that the first document d_1 is "closer" to the query q when the distance is defined as the angle made between the corresponding vectors. The second document d_2 is "closer" to the query q when the distance is measured using the Euclidean norm. That is, $\cos \theta_1 < \cos \theta_2$, while $\|d_1 - q\|_2 > \|d_2 - q\|_2$.

Each query is modeled as a vector using the same attribute space as the documents. The relevancy ranking of a document with respect to a query depends on its *distance* to the query vector. In our experiments, we use the cosine of the angle defined by the query and document vectors as the distance metric, because it is relatively simple to compute, and implementation experiments have indicated that it tends to lead to better IR results than the Euclidean distance (see Figure 5.2). Numerous other distances have been proposed for various applications and database sizes [Har99].

Many databases are so massive that the similarity ranking method described above requires too many computations and comparisons for real-time output. Indeed, scalability of relevancy ranking methods to massive databases is a serious concern as users consistently select the most important feature of IR engines to be fast real-time response to their queries.[4] One approach for solving this problem is to reduce the dimension of mathematical models by projection into a subspace

---

[3]Some researchers believe that this basic assumption of correlation between term frequency and its importance is not valid. For example, when a document is about a single main subject, it need not be explicitly mentioned in every sentence since it is tacitly understood. Consequently, its term frequency may be quite low relative to its importance. However, if a document is about multiple subjects, each time a particular person or object is described, it must be explicitly mentioned to avoid ambiguity, so the term frequencies of all of the subjects will be very high. Further examples and detailed discussion are given in [Kat96].

[4]Graphics, Visualization, and Usability Center of Georgia Institute of Technology (GVU) Web users' survey: http://www.gvu.gatech.edu/user_surveys.

of sufficiently small dimension to enable fast response times, but large enough to retain characteristics for distinguishing contents of individual documents. In this chapter we consider two algorithms for carrying out dimensional reduction: *latent semantic indexing* and a variation of *principal component analysis*. We note that at least two other approaches for reducing the dimension of VSM-based IR have been proposed: centroid and least squares analysis by Park, Jeon, and Rosen [PJR01, PJR03] and a Krylov subspace method by Blom and Ruhe [BR01].

5.2.2   Latent Semantic Indexing (LSI)

Suppose we are given a database with m documents and n distinguishing attributes for relevancy ranking. Let A denote the corresponding $m \times n$ document-attribute matrix model with entries $a(i, j)$ that represent the importance of the ith term in the jth document. The fundamental idea in LSI is to reduce the dimension of the IR problem to k, where $k \ll m, n$, by projecting the problem into the space spanned by the rows of the closest rank-k matrix to A in the Frobenius norm [DDF+90]. Projection is performed by computing the singular value decomposition of A [GV96], then constructing a modified matrix A_k from the k-largest singular values $\sigma_i, i = 1, 2, 3, \ldots, k$, and their corresponding vectors:

$$A_k = U_k \, \Sigma_k \, V_k^T .$$

Σ_k is a diagonal matrix with monotonically decreasing diagonal elements σ_i. The columns of matrices U_k and V_k are the left and right singular vectors of the k-largest singular values of A.

Processing a query takes place in two steps: projection, followed by matching. In the projection step, input queries are mapped to pseudo-documents in the reduced query-document space by the matrix U_k, then weighted by the inverses of the corresponding singular values σ_i (diagonal elements of the matrix Σ_k):

$$q \longrightarrow \hat{q} = q^T \, U_k \, \Sigma_k^{-1} ,$$

where q represents the original query vector, and $\hat{q}$ is the pseudo-document. In the second step, similarities between the pseudo-document $\hat{q}$ and documents in the reduced term document space V_k^T are computed using a similarity measure, such as the angle defined by a document and query vector.

The inventors of LSI claim that the dimensional reduction process reduces unwanted information or "noise" in the database, and aids in overcoming synonymy and polysemy problems. *Synonymy* refers to the existence of equivalent or similar terms that can be used to express an idea or object in most languages, and *polysemy* refers to the fact that some words have multiple unrelated meanings [DDF+90]. Absence of accounting for synonymy will lead to many small disjoint clusters, some of which should actually be clustered, whereas absence of accounting for polysemy can lead to clustering of unrelated documents.

A major bottleneck in applying LSI to massive databases is efficient and accurate computation of the largest few hundred singular values and the corresponding singular vectors of the highly sparse document-query matrix. Even though document-attribute matrices that appear in IR tend to be very sparse (usually 0.2 to 0.3% nonzero), computation of the top 200 to 300 singular triplets of the matrix using a powerful desktop PC becomes impossible when the number of documents exceeds several hundred thousand.

5.2.3   *Covariance Matrix Analysis (COV)*

The scalability issue associated with LSI can be resolved in many practical applications in a dimensional reduction method known as *principal component analysis*, invented first by Pearson [Pea01] in 1901 and independently reinvented by Hotelling [Hot33] in 1933. PCA has several different names, depending on the context in which it is used. It has also been referred to as *the Kahrhunen–Loève procedure*, *eigenvector analysis*, and *empirical orthogonal functions*. Until recently it has been used primarily in statistical data analysis and image processing.

We review a PCA-based algorithm for text and data mining that focuses on COVariance matrix analysis, or *COV*. In the COV algorithm, document and query vectors are projected onto the subspace spanned by the k eigenvectors corresponding to the k-largest eigenvalues of the covariance matrix of the document vectors C. In other words, the IR problem is mapped to a subspace spanned by a subset of the principal components that correspond to the k-highest principal values. Stated more rigorously, given a database modeled by an $m \times n$ document-attribute term matrix A, with m row vectors $\{d_i^T \mid i = 1, 2, 3, \ldots, m\}$ representing documents, each having n dimensions representing attributes, the covariance matrix for the set of document vectors is

$$C \equiv \frac{1}{M} \sum_{i=1}^{M} d_i d_i^T - \bar{d}\,\bar{d}^T \, , \tag{5.1}$$

where d_i represents the ith document vector and $\bar{d}$ is the componentwise average over the set of all document vectors [MKB79]. That is,

$$\bar{d} = [\bar{a}_1 \; \bar{a}_2 \; \bar{a}_3 \; \cdots \; \bar{a}_n]^T \, ;$$
$$d_i = [a_{i,1} \; a_{i,2} \; a_{i,3} \; \cdots \; a_{i,n}]^T \, ;$$
$$\bar{a}_j = \frac{1}{m} \sum_{i=1}^{m} a_{i,j} \, .$$

Since the covariance matrix is symmetric positive semidefinite, it can be decomposed into the product

$$C = V \, \Sigma \, V^T \, ,$$

where V is an orthogonal matrix that diagonalizes C so that the diagonal entries of Σ are in monotone decreasing order going from top to bottom, that is, diag(Σ) = $(\lambda_1, \lambda_2, \lambda_3, \ldots, \lambda_n)$, where $\lambda_i \geq \lambda_{i+1}$ for $i = 1, 2, 3, \ldots, n$ [Par97]. To reduce the dimension of the IR problem to $k \ll m, n$, we project all document vectors and the query vector into the subspace spanned by the k eigenvectors $\{v_1, v_2, v_3, \ldots, v_k\}$

corresponding to the k-largest eigenvalues $\{\lambda_1, \lambda_2, \lambda_3, \ldots, \lambda_k\}$ of the covariance matrix C. Relevancy ranking with respect to the modified query and document vectors is performed in a manner analogous to the LSI algorithm, that is, projection of the query vector into the k-dimensional subspace followed by measurement of the similarity.

5.2.4 Comparison of LSI and COV

Implementation studies for IR have shown that LSI and COV are similar in that they both project a high-dimensional problem into a subspace to reduce computational requirements for determining relevancy rankings of documents with respect to a query, but large enough to retain distinguishing characteristics of documents to enable accurate retrieval [KAST02].

However, the algorithms differ in several ways. For instance, they use different criteria to determine a subspace: LSI finds the space spanned by the rows of the closest rank-k matrix to A in the Frobenius norm [EY39], while COV finds the k-dimensional subspace that best represents the full data with respect to the minimum square error [KMS00]. COV shifts the origin of the coordinate system to the "center" of the subspace to spread documents apart as much as possible so that documents can more easily be distinguished from one another (as shown in Figure 5.3).

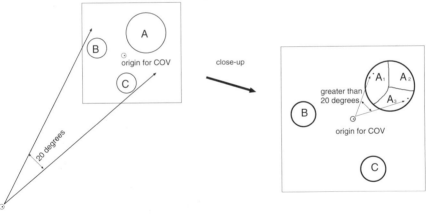

Figure 5.3. LSI and COV map the relevancy ranking problem into a proper subspace of the document-attribute space. While LSI does not move the origin, COV shifts the origin to the "center" of the set of basis vectors in the subspace so that document vectors are more evenly spaced apart, enabling finer detection of differences between document vectors in different clusters and subclusters.

A second advantage of the COV algorithm is scalability. The primary bottleneck of COV is the computation of the largest few hundred eigenvalues and corresponding eigenvectors of a square, symmetric, positive semidefinite matrix of dimension less than or equal to the dimension of the attribute space [KMS00]. Because the dimension of a covariance matrix is independent of the number of documents, COV can be used for real-time IR and data mining as long as the attribute space is relatively small. Usually the dimension of the attribute space is less than 20,000, so computations can be performed in the main memory of a standard PC. When the dimension of the attribute space is so large that the covariance matrix cannot fit into the main memory, the eigenvalues of the covariance matrix can be computed using an implicit construction of the covariance matrix [KMS00]. This implicit covariance method is generally much slower, especially for databases with hundreds of thousands of documents since paging will occur.

Neural networks pose another alternative for estimating the eigenvalues and eigenvectors of the covariance matrix [Hay99]. The advantage of the neural network approach is that it tends to be less computationally expensive. However, the disadvantages make the method unattractive for IR. For example, it is not clear when convergence to an eigenvector has occurred (although convergence can be tested, to some extent, by multiplying the vector to the matrix, and examining the difference in the angle made by the input and output vectors). Also, it is not clear which eigenvector has been found, that is, whether it is the one corresponding to the largest, smallest, or kth largest eigenvalue. In addition, the neural network may not converge at all. In short, neural network approaches are not suitable since there is no guaranteed means by which the eigenvectors corresponding to the k-largest eigenvalues can be computed.

A third attractive feature of COV is an algorithm developed by Qu, Ostrou-chov, Samatova, and Geist [QOSG02] for mining information from datasets that are distributed across multiple locations. The main idea of the algorithm is to compute *local* principal components for dimensional reduction for each location. Information about local principal components is subsequently sent to a centralized location, and used to compute estimates for the *global* principal components. The advantage of this method over a centralized (nondistributed) approach and a parallel processing approach is the savings in data transmission rates. Data transmission costs often exceed computational costs for large data sets [Dem97]. More specifically, transmission rates will be of order $O(sp)$ instead of $O(np)$, where n is the total number of all documents over all locations, p is the number of attributes, and s is the number of locations. According to the authors, when the dominant principal components provide a good representation of the datasets, the algorithm can be as equally accurate as its centralized (nondistributed) counterpart in implementation experiments. However, if the dominant components do not provide a good representation, up to 33% more components need to be computed to attain a level of accuracy comparable to its centralized counterpart so the subspace into which the IR problem will be mapped will be significantly larger.

5.3   VSM for Major and Minor Cluster Discovery

5.3.1   Clustering

Another approach for overcoming the scalability issue with IR and data mining systems is to identify *clusters* (or sets of documents that cover similar topics) during the preprocessing stage so they can be retrieved together to reduce the query response time [Ras92]. This approach, which has come to be known as *cluster retrieval*, is based on the premise known as the *cluster hypothesis* that "closely associated documents tend to be relevant to the same requests." In cluster retrieval, only documents in one or a few selected clusters are ranked and presented to users.

We note that the study of clusters in large databases encompasses a much broader spectrum of ideas and applications, all of which have to do with the development of meaningful structures and taxonomies for an intended use. For example, cluster identification and analysis can also be used to understand issues and trends addressed by documents in massive databases.

In this section, we focus on LSI- and COV-based methods for cluster analysis, although numerous other approaches are available in the literature [De02, JD88]. One of the attractive features of both methods is the recognition and preservation of naturally occurring overlaps of clusters. Preservation of overlap information facilitates multiperspective analysis of database contents. Most cluster analysis methods partition sets of documents so that cluster overlaps are not permitted and may lead to distorted results. For example, if two clusters X and Y have significant overlap, and the number of documents in $Y/(X \cap Y)$ (the set of documents in Y that do not belong to $X \cap Y$) is very small, a mining system may report the existence of cluster X, and discard documents in $Y/(X \cap Y)$. To summarize, identification of naturally occurring overlap of clusters may prevent omission of clusters that have a significant overlap with others.

LSI and COV can be used to find *major* document clusters by using basis vectors from the reduced dimensional space to identify major topics comprised of sets of several keywords. Implementation studies show that results from the two algorithms are usually very close [KA02]. Both algorithms are not as successful at finding smaller, *minor* document clusters, because major clusters and their large subclusters dominate the subspace basis vector selection process. And during the dimensional reduction step in LSI and COV, documents in minor clusters are often mistaken for noise and are discarded. Mining information in minor clusters in massive databases represents one of the current big challenges in data mining research, because minor clusters may represent valuable nuggets of information [SY02].

5.3.2   Iterative Rescaling: Ando's Algorithm

Recently, Ando [And00] proposed an algorithm to identify small clusters in limited contexts. Her implementation studies were carried out using very small document sets consisting of 684 TREC documents.[5] The main intended idea in her algorithm was to prevent major themes from dominating the process of selecting the basis vectors for the subspace projection step in LSI. This is supposed to be carried out during the basis vector selection process by introducing an unfavorable weight (also known as filter or negative bias) to documents that belong to clusters that are well represented by basis vectors that have already been selected. The weight is imparted by computing the magnitude (i.e., the length in the Euclidean norm) of the *residual* of each document vector (i.e., the proportion of the document vector that cannot be represented by the basis vectors that have been selected up to that point), then rescaling the magnitude of each document vector by a power q of the magnitude of its residual.

Algorithm 5.1 Ando's Algorithm

$R = A$;
for $(i = 1; i \leq k; i + +)\{$
    $R_s = [|r_1|^q r_1, |r_2|^q r_2, |r_3|^q r_3, \ldots, |r_m|^q r_m]^T$;
    $b_i = $ the first eigenvector of $R_s^T R_s$;
    $R = R - R b_i b_i^T$;   (the residual matrix)
$\}$

The input parameters are the document-term matrix A, the constant scale factor q, and the dimension k to which the IR problem will be mapped. The *residual matrices* are denoted by R and R_s. We set R to be A initially. After each iterative step the residual vectors are updated to take into account the new basis vector b_i. After the kth basis vector is computed, each document vector d_j in the original IR problem is mapped to its counterpart $\hat{d}_j$ in the k-dimensional subspace as follows, $\hat{d}_j = [b_1, b_2, b_3, \ldots, b_k]^T d_j$. The query vector is mapped to the k-dimensional subspace before relevance ranking is performed.

Ando's algorithm encounters the following problems during implementations.

- All major and all minor clusters may not be identified;

- the procedure for finding eigenvectors may become unstable when the scaling factor q is large;

- the basis vectors b_i are not always orthogonal; and

- if the number of documents in the database is even moderately large, intermediate data will not fit into main memory on an ordinary PC after a few iterative steps.

The first two points can be explained as follows. The algorithm usually fails to yield good results because rescaling document vectors after computation of each basis

---

[5]The full dataset was used in the Text REtrieval Competition (TREC) sponsored by the United States National Institute of Standards and Technology: http://trec.nist.gov.

vector leads to the rapid diminution of documents that have even a moderate-size component in the direction of one of the first few document vectors. To understand how weighting can obliterate these vectors, consider the following scenario. Suppose that a document has a residual of 90% after one basis vector is computed, and q is set to be one. Before the next iteration, the vector is rescaled to length 0.81, after two more iterations $0.81 \times 0.81 < 0.66$, and after n more iterations, less than 0.81 to the nth power.

The third point appears critical at first, particularly from a theoretical perspective. However, implementation studies of IR using news databases indicate that slight deviations from orthogonality appear to cause only small differences in relevancies but no differences in the relevancy rankings among documents [KAST02]. In fact, the simplest solution for correcting nonorthogonality of the basis vectors would be to introduce Gram–Schmidt orthogonalization [GV96], but the significant increase in computational cost appears to be unjustifiable considering the very small changes (if any) in the ranking results.

The fourth and final point regarding the scalability of the algorithm is crucial to data mining applications. Document-attribute matrices are usually highly sparse, typically having only 0.2 to 0.3% nonzero elements. After only a few iterative steps of the algorithm, the document-attribute (residual) matrix will become completely dense and will not fit in the main memory of a good PC if the database under consideration is only moderately large.

More recently, Ando and Lee [AL01] proposed a refinement of Ando's earlier algorithm, given above. The major new contribution has been to automatically set the rescaling factor. Implementation studies were conducted using a 684 document subset from the TREC collection. The new algorithm still encounters the scalability issue mentioned above.

In the next two subsections, we propose two new algorithms for identifying major and minor clusters that overcome some of the problems associated with Ando's algorithm. The first is a significant modification of LSI and Ando's algorithm and the second is based on COV.

5.3.3   Dynamic Rescaling of LSI

The first algorithm we propose is based on LSI and can only be applied to small databases. It was developed independently by Kobayashi et al. [KAST01] at about the same period as that by Ando and Lee [AL01]. The algorithm by Kobayashi et al. encounters the same difficulties in scalability as the earlier and more recent algorithms by Ando and Lee. The second algorithm that is based on COV is scalable to large databases. Like Ando's algorithm, the idea in both of our algorithms is to prevent major themes from dominating the process of selecting basis vectors for the subspace into which the IR problem will be projected. This is carried out by introducing weighting to decrease the relative importance of attributes that are already well represented by subspace basis vectors that have already been computed. The weighting is dynamically controlled to prevent deletion of information in both major and minor clusters.

Algorithm 5.2 LSI-rescale (minor cluster mining based on LSI and rescaling)

---

for ($i = 1; i \leq k; i + +$){

   $t_{max} = \max(|r_1|, |r_2|, |r_3|, \ldots, |r_m|)$;

   $q = \text{func}(t_{max})$;

   $R_s = \left[|r_1|^q \, r_1, \; |r_2|^q \, r_2, \; |r_3|^q \, r_3, \ldots, \; |r_m|^q \, r_m \right]^T$;

   SVD (R_s) ;   (the singular value decomposition)

   $\hat{b}_i = $ the first row vector of V^T ;

   $b_i = \text{MGS}(\hat{b}_i)$;   (modified Gram–Schmidt)

   $R = R - R \, b_i b_i^T$;   (residual matrix)

}

The input parameters for this algorithm are the document-term matrix A, the re-scaling factor q (used for weighting), and the dimension k to which the problem will be reduced. The *residual matrices* are denoted by R and R_s. Initially, R is set to be A. After each iterative step the residual vectors are updated to take into account the most recently computed basis vector b_i. After the kth basis vector is computed, each document vector d_j in the original IR problem is mapped to its counterpart $\hat{d}_j$ in the k-dimensional subspace: $\hat{d}_j = [b_1, b_2, b_3, \ldots, b_k]^T d_j$. The query vector is mapped to the k-dimensional subspace before relevance ranking is performed.

The LSI-rescale algorithm is based on the observation that rescaling document vectors after computation of each basis vector can be useful, but the weighting factor should be reevaluated after each iterative step to take into account the length of the residual vectors to prevent decimation from overweighting (or overreduction). More specifically, in the first step of the iteration we compute the maximum length of the residual vectors and use it to define the scaling factor q that appears in the second step.

$$q = \begin{cases} t_{max}^{-1} & \text{if} \quad t_{max} > 1 \\ 1 + t_{max} & \text{if} \quad t_{max} \approx 1 \\ 10^{t_{max}^{-2}} & \text{if} \quad t_{max} < 1 \end{cases} .$$

5.3.4   Dynamic Rescaling of COV

The second algorithm, COV-rescale, for minor cluster identification is a modification of COV, analogous to LSI-rescale and LSI. In COV-rescale, the residual of the covariance matrix (defined by Eq. (5.1)) is computed. Results from our implementation studies (given in the next section) indicate that COV-rescale is better than LSI, COV, and LSI-rescale at identifying large and multiple minor clusters.

The input parameters for this algorithm are the covariance matrix C for the document residual vectors (given by Eq. (5.1)) the rescaling factor q (used for weighting), and the dimension k to which the problem will be reduced. The *residual matrices* are denoted by R and R_s. Initially, R is set to be the covariance matrix for the set of document vectors.

Algorithm 5.3 COV-rescale (minor cluster mining based on rescaling and COV)

for $(i = 1; i \le k; i++)\{$
  $t_{max} = \max(|r_1|, |r_2|, |r_3|, \ldots, |r_m|)$;
  $q = \text{func}(t_{max})$;
  $R_s = [|r_1|^q r_1, |r_2|^q r_2, |r_3|^q r_3, \ldots, |r_m|^q r_m]^T$;
  $C = \text{COV}(R_s)$;   (covariance matrix)
  $\text{SVD}(C)$;   (singular value decomposition)
  $\hat{b}_i = $ the first row vector of V^T ;
  $b_i = \text{MGS}(\hat{b}_i)$;   (modified Gram–Schmidt)
  $R = R - R\, b_i b_i^T$;   (residual matrix)
}

5.4   Implementation Studies

To test and compare the quality of results from the algorithms discussed above, we performed several sets of numerical experiments: the first with a small data set to test our concept, the second with a larger data set to demonstrate the applicability of our method to realistic size data, and the third with an *L.A. Times* news article dataset used in TREC competitions.

5.4.1   Implementations with Artificially Generated Datasets

For our first set of experiments we constructed a data set consisting of two large major clusters (each of which had three subclusters), four minor clusters, and noise. Each major cluster had two subclusters that were twice as large as the minor clusters and a subcluster that was the same size as the minor clusters, as shown below.

Cluster Structure of Small Dataset (140 documents, 40 terms)
  25 docs (Clinton cluster) – *major cluster*
    10 docs (Clinton + Gore) – *subcluster*
    10 docs (Clinton + Hillary) – *subcluster*
    10 docs (Clinton + Gore + Hillary) – *subcluster*
  25 docs (Java cluster) – *major cluster*
    10 docs (Java + JSP) – *subcluster*
    5 docs (Java + Applet) – *subcluster*
    10 docs (Java + JSP + Applet) – *subcluster*
  5 docs (Bluetooth cluster) – *minor cluster*
  5 docs (Soccer cluster) – *minor cluster*
  5 docs (Matrix cluster) – *minor cluster*
  5 docs (DNA cluster) – *minor cluster*
  70 docs *noise*

We implemented five algorithms to reduce the dimension of the document-term space from 40 dimensions to 6 dimensions: LSI, COV, Ando, LSI-rescale, and COV-rescale. Table 5.1 summarizes clusters that were detected as basis vectors were computed (uppercase letters are used to label major clusters and lowercase for minor clusters and noise). **C** represents the major cluster *Clinton*, **J** the major cluster *Java*, **noise** *noise*, **b** the minor cluster *Bluetooth*, **s** the minor cluster *soccer*, **m** the minor cluster *matrix*, **d** the minor cluster *DNA*, and **all-m** the set of all minor clusters.

LSI failed to detect any information in minor clusters. COV picked up some information in minor clusters, but failed to detect specific minor clusters. Ando's algorithm detected two minor clusters: *Bluetooth* and *soccer* in b_4 and the two remaining minor clusters *matrix* and *DNA* in b_5 and b_6. Our results indicate that after the fourth iteration the lengths of the residual vectors for documents covering topics other than *matrix* and *DNA* have been given too much unfavorable weighting so information in them cannot be recovered. Furthermore, it demonstrates why re-scaling using a constant factor q does not work well when there are multiple minor clusters. In contrast, LSI-rescale and COV-rescale successfully detect all major and minor clusters.

Results from using LSI-rescale are as follows: *matrix* and *DNA* are detected by b_4; *Bluetooth* and *soccer* by b_5; and *minor* (the set of all minor clusters) by b_6. In short, all minor clusters are detected evenly. Results for COV-rescale are: *matrix* and *DNA* are detected by b_3; *Bluetooth* and *soccer* by b_5; and *minor* by b_2, b_4, and b_6; that is, all minor clusters are evenly detected, as in LSI-rescale.

Vector	LSI	COV	Ando	LSI-Rescale	COV-Rescale
b_1	C	C, J	J	J	C, J
b_2	J	C, J	C	C	noise, all-m, C, J
b_3	noise	noise, all-m	noise	noise	m, d
b_4	C	all-m, noise	b, S	M, D	all-m
b_5	J	all-m, noise	m, d	b, s	b, s
b_6	noise	all-m, noise	m, d	all-m	noise, all-m

Table 5.1. Clusters Identified Using Subspace Basis Vectors b_i in First Experiment

For our second set of experiments we constructed a data set consisting of 10,000 documents and 500 attributes (keywords) with 5 major clusters *M1, M2, M3, . . . , M5* (each with 500 documents or 5% of the total), 20 minor clusters *m1, m2, m3, . . . , m20* (each with 200 documents or 2% of the total), and 35% noise. We performed numerical experiments to reduce the dimension of the document-attribute space to 25 dimensions and observed how many major and minor clusters were detected.

For the straightforward LSI algorithm, computation of 25 basis vectors took 0.96 seconds, for COV 1.05 seconds, and for COV with iterative rescaling (COV-rescale) 263 seconds. Since the computation time for LSI did not include conversion into Harwell–Boeing format, and the computation times for the COV algorithms include construction of the covariance matrix, the times for LSI and basic COV are comparable. Iterative rescaling is very expensive for COV because an updated covariance matrix must be constructed after each set of computations to find one eigenvector. However, our numerical experiments show that the extra computation time is worth the effort if the user is interested in finding as many major and minor clusters as possible. LSI detects only 19 clusters (5 major + 14 minor), basic COV detects only 21 (5 major + 16 minor), while COV with rescaling detects all 25 major and minor clusters. Our results are summarized in Table 5.2. Finally, we note that an implementation of LSI with rescaling (Ando's Algorithm and LSI-rescale) would have been too difficult for this dataset because after a few steps of rescaling, we would have been required to compute the SVD of a very large, dense matrix.

Vector	LSI	COV	COV-Rescale
b_1	M1	M3	M1, M3
b_2	M5	M1	M2, M4
b_3	M4	M5	m12, m17
b_4	M2	M4	m14, m17
b_5	M3	M2	m4, m14
b_6	M1	M3	m1, m14
b_7	M5	M1	m11, m18
b_8	M4	M5	m18
b_9	M2	M4	m6
b_{10}	M3	M2	m7, m16
b_{11}	m4	m8, m19	m7, m20
b_{12}	m2	m7, 19	m10, m20
b_{13}	M2	m7	m2, m10
b_{14}	m19	m12, m16	m13
b_{15}	m16	m12, m13	m8, m13
b_{16}	m14	m13, m15	m3, m8
b_{17}	m12	m3, m15	m3, m11
b_{18}	m5	m3, m11	m11, m19
b_{19}	m13	m9, m11	m9, m19
b_{20}	m20	m9, m14	m5
b_{21}	m17	m5, m14	m15
b_{22}	m10	m5, m20	M5
b_{23}	m7	m1, m20	M5
b_{24}	m15	m1, m17	M4
b_{25}	m3	m2, m17	*noise*

Table 5.2. Clusters Identified Using Subspace Basis Vectors b_i in Second Experiment

5.4.2   Implementations with L.A. Times News Articles

In this subsection we describe results from experiments with an *L.A. Times* database to identify major and minor clusters using the COV-rescale algorithm presented above. The *L.A. Times* electronic database consists of more than 100,000 news articles provided by the owners to the TREC participants. The documents were converted to vector format using an in-house stemming tool[6] based on natural language processing technologies, followed by manual deletion of stop-words and words that appear infrequently. A total of 11,880 terms were selected to be the document attributes. The *TF-IDF* model described in Section 5.2.1 was used for term-weighting. In the final preprocessing step, the COV method was used to reduce the dimension of the IR-clustering problem to 200.

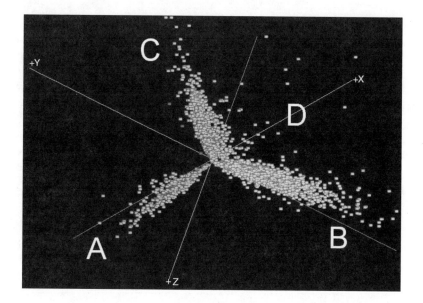

Figure 5.4. Visualization of three major clusters in the *L.A. Times* news database when document vectors are projected into the 3-D subspace spanned by the three most relevant axes (basis vectors 1, 2, and 3) determined using COV rescale. The major clusters (marked A, B, and C) are comprised of articles about {*Bush, US, Soviet, Iraq*}, {*team, coach, league, inning*}, and {*police, digest, county, officer*}, respectively. A faint trace of a minor cluster (marked D) on {*Zurich, Swiss, London, Switzerland*} can be seen in the background.

Figure 5.4 shows a screen image from our prototype cluster mining system. The coordinate axes for the subspace are the first three basis vectors output by the COV rescale algorithm. The three major clusters, marked A, B, and C, are comprised of

---

[6]Owned by IBM Research.

articles about {*Bush, US, Soviet, Iraq*}, {*team, coach, league, inning*}, and {*police, digest, county, officer*}, respectively. A faint trace of a minor cluster (marked D) on {*Zurich, Swiss, London, Switzerland*} can be seen in the background. This minor cluster can be seen much more clearly when other coordinate axes are used for visualization and display. This example shows that careful analysis is needed before deciding whether a faint cloud of points is noise or part of a cluster.

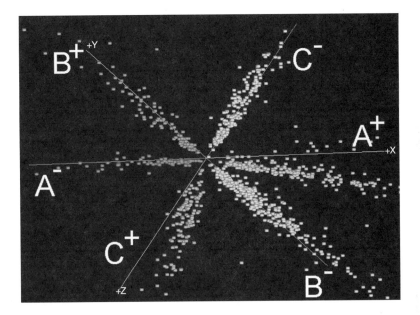

Figure 5.5. Visualization of six minor clusters in the L.A. Times news database by our prototype cluster mining system when document vectors are projected into the 3-D space spanned by basis vectors 58, 63, and 104.

We continued computation of more basis vectors that could be used to view three-dimensional slices. After many major clusters were found, minor clusters began to appear. An example of a screen image that displays six minor clusters using basis vectors 58, 63, and 104 is given in Figure 5.5. Note that two clusters may lie along a coordinate axis, one each in the positive and negative directions. The clusters are comprised of articles on {abortion, anti-abortion, clinic, Roe}, {lottery, jackpot, California, ticket}, {AIDS, disease, virus, patient}, {gang, school, youth, murder}, {Cypress, Santiago, team, tournament}, and {jazz, pianist, festival, saxophonist}, respectively. A plus or minus sign is used to mark the direction in which a cluster lies.

5.5   Conclusions and Future Work

We presented a new algorithm based on covariance matrix analysis for mining clusters in very large databases. The novel features of our algorithm are:

- high scalability,

- ability to identify major as well as minor clusters, and

- accomodation of cluster overlaps.

We implemented our algorithm using artificially generated data with known cluster structure. Our experimental results agreed with the correct (expected) answer. We subsequently conducted experiments using a "real" database with over 100,000 documents from the TREC competition. We found that our algorithm based on covariance matrix analysis requires fewer computations than other proposed algorithms for minor cluster mining that are based on the singular value decomposition, because clusters can be identified along both the positive and negative directions of the subspace basis vector axes in covariance-based methods, while clusters can only be identifed in the positive direction along the subspace basis vector axes in singular value decomposition-based methods.

In the near future, we plan to investigate and refine speed-up methods for the preprocessing steps of our algorithm and the search engine component of our system.

Acknowledgments: The authors would like to thank Hikaru Samukawa, Michael Houle, Hironori Takeuchi, and Kohichi Takeda for many helpful conversations and Eric Brown for providing us access to the TREC dataset. This study was supported by the Research Division of IBM Corporation.

References

[AL01]   R. Ando and L. Lee.Latent semantic space.In *Proceedings of the ACM Special Interest Group for Information Retrieval (SIGIR) Conference*, Helsinki, Finland, pages 154–162, 2001.

[And00]   R. Ando.Latent semantic space.In *Proceedings of the ACM Special Interest Group for Information Retrieval (SIGIR) Conference*, Athens, pages 216–223, 2000.

[BDJ99]   M. Berry, Z. Drmač, and E. Jessup.Matrices, vector spaces, and information retrieval.*SIAM Review*, 41(2):335–362, 1999.

[BDO95]   M. Berry, S. Dumais, and G. O'Brien.Using linear algebra for intelligent information retrieval.*SIAM Review*, 37(4):573–595, 1995.

[BR01]   K. Blom and A. Ruhe.Information retrieval using very short Krylov sequences.In *Proceedings of the Computational Information Retrieval Confer-*

ence held at North Carolina State University, Raleigh, Oct. 22, 2000, M. Berry, ed., SIAM, Philadelphia, pages 39–52, 2001.

[DDF+90] S. Deerwester, S. Dumais, G. Furnas, T. Landauer, and R. Harshman.Indexing by latent semantic analysis.*Journal of the American Society for Information Science*, 41(6):391–407, 1990.

[De02] I. Dhillon and J. Kogan (eds.).*Proceedings of the Workshop on Clustering High Dimensional Data and its Applications*.SIAM, Philadelphia, 2002.

[Dem97] J. Demmel.*Applied Numerical Linear Algebra*.SIAM, Philadelphia, 1997.

[EY39] C. Eckart and G. Young.A principal axis transformation for non-Hermitian matrices.*Bulletin of the American Mathematics Society*, 45:118–121, 1939.

[GV96] G. Golub and C. Van Loan.*Matrix Computations*, third edition.John Hopkins Univ. Press, Baltimore, MD, 1996.

[Har99] D. Harman.Ranking algorithms.In *Information Retrieval*, R. Baeza-Yates and B. Ribeiro-Neto, eds., ACM, New York, pages 363–392, 1999.

[Hay99] S. Haykin.*Neural Networks: A comprehensive foundation, second edition.* Prentice-Hall, Upper Saddle River, NJ, 1999.

[Hot33] H. Hotelling.Analysis of a complex of statistical variables into principal components.*Journal of Educational Psychology*, 24:417–441, 1933.

[JD88] A. Jain and R. Dubes.*Algorithms for Clustering Data*.Prentice-Hall, Englewood Cliffs, NJ, 1988.

[KA02] M. Kobayashi and M. Aono.Major and outlier cluster analysis using dynamic re-scaling of document vectors.In *Proceedings of the SIAM Text Mining Workshop, Arlington, VA*, SIAM, Philadelphia, pages 103–113, 2002.

[KAST01] M. Kobayashi, M. Aono, H. Samukawa, and H. Takeuchi.*Information retrieval apparatus for accurately detecting multiple outlier clusters*.patent, filing, IBM Corporation, 2001.

[KAST02] M. Kobayashi, M. Aono, H. Samukawa, and H. Takeuchi.Matrix computations for information retrieval and major and outlier cluster detection.*Journal of Computational and Applied Mathematics*, 149(1):119–129, 2002.

[Kat96] S. Katz.Distribution of context words and phrases in text and language modeling.*Natural Language Engineering*, 2(1):15–59, 1996.

[KMS00] M. Kobayashi, L. Malassis, and H. Samukawa.*Retrieval and ranking of documents from a database*.patent, filing, IBM Corporation, 2000.

[MKB79] K. Mardia, J. Kent, and J. Bibby.*Multivariate Analysis*.Academic, New York, 1979.

[MS00] C. Manning and H. Schütze.*Foundations of Statistical Natural Language Processing*.MIT Press, Cambridge, MA, 2000.

[Par97] B. Parlett.*The Symmetric Eigenvalue Problem*.SIAM, Philadelphia, 1997.

[Pea01] K. Pearson.On lines and planes of closest fit to systems of points in space.*The London, Edinburgh and Dublin Philosophical Magazine and Journal of Science, Sixth Series*, 2:559–572, 1901.

[PJR01] H. Park, M. Jeon, and J. Rosen.Lower dimensional representation of text data in vector space based information retrieval.In *Proceedings of the Computational Information Retrieval Conference held at North Carolina State University, Raleigh, Oct. 22, 2000*, M. Berry, ed., SIAM, Philadelphia, pages 3–24, 2001.

[PJR03]   H. Park, M. Jeon, and J.B. Rosen.Lower dimensional representation of text data based on centroids and least squares.*BIT*, 2003, to appear.

[QOSG02]   Y. Qu, G. Ostrouchov, N. Samatova, and A. Geist.Principal component analysis for dimension reduction in massive distributed data sets.In *SIAM Workshop on High Performance Data Mining*, S. Parthasarathy, H. Kargupta, V. Kumar, D. Skillicorn, and M. Zaki, eds., Arlington, VA, pages 7–18, 2002.

[Ras92]   E. Rasmussen.Clustering algorithms.In *Information Retrieval*, W. Frakes and R. Baeza-Yates, eds., Prentice-Hall, Englewood Cliffs, NJ, pages 419–442, 1992.

[Sal71]   G. Salton.*The SMART Retrieval System*.Prentice-Hall, Englewood Cliffs, NJ, 1971.

[SY02]   H. Sakano and K. Yamada.Horror story: The curse of dimensionality.*Information Processing Society of Japan (IPSJ) Magazine*, 43(5):562–567, 2002.

[WF99]   I. Witten and E. Frank.*Data Mining: Practical Machine Learning Tools and Techniques with Java Implementations*.Morgan Kaufmann, San Francisco, 1999.

6

HotMiner: Discovering Hot Topics from Dirty Text

Malú Castellanos

Overview

For companies with websites that contain millions of documents available to their customers, it is critical to identify their customers' hottest information needs along with their associated documents. This valuable information gives companies the potential of reducing costs and being more competitive and responsive to their customers' needs. In particular, technical support centers could drastically lower the number of support engineers by knowing the topics of their customers' hot problems (i.e., *hot topics*), and making them available on their websites along with links to the corresponding solutions documents so that customers could efficiently find the right documents to self-solve their problems. In this chapter we present a novel approach to discovering hot topics of customers' problems by mining the logs of customer support centers. Our technique for *search log* mining discovers hot topics that match the user's perspective, which often is different from the topics derived from document content categorization[1] methods. Our techniques to mine *case logs* include extracting relevant sentences from cases to conform case excerpts which are more suitable for hot topics mining. In contrast to most text mining work, our approach deals with *dirty text* containing typos, adhoc abbreviations, special symbols, incorrect use of English grammar, cryptic tables and ambiguous and missing punctuation. It includes a variety of techniques that either directly handle some of these anomalies or that are robust in spite of them. In particular, we have developed a postfiltering technique to deal with the effects of noisy clickstreams due to random clicking behavior, a Thesaurus Assistant to help in the generation of a thesaurus of "dirty" variations of words that is used to normalize the terminology, and a Sentence Identifier with the capability of eliminating code and tables. The techniques that compose our approach have been implemented as a toolbox,

---

[1] *Categorization* here means the discovery of categories from a collection of documents along with the placement of such documents into those categories.

HotMiner, which has been used in experiments on logs from Hewlett-Packard's Customer Support Center.

6.1   Introduction

The information that organizations make available on their websites has become a key resource for satisfying their customers' information needs. However, organizing the contents of a website with millions of documents into topics is not an easy task. The organization of the site has to be such that it allows customers to do searches and navigation efficiently. Customers who have to spend a lot of time trying to find the information that corresponds to their needs get frustrated and are unlikely to return to the site. Thus the site becomes a potential source of disgruntled customers, instead of a source to attract and keep loyal customers.

Categorizing documents into a topic hierarchy has traditionally been done according to the document contents, either manually or automatically. The manual approach has been the most successful so far. In Yahoo!, for example, topics are obtained manually. However, even though the documents in Yahoo! are correctly categorized and the categories (topics) are natural and intuitive, the manual effort that is required to do the categorization is huge and can only be done by domain experts. In response to these requirements, many efforts have been dedicated to creating methods that mine the contents of a document collection to do the categorization automatically. Unfortunately, so far none of the automatic approaches has obtained results comparable to the ones obtained by manual categorization; that is, the resulting categories are not always intuitive or natural. However, if instead of mining the contents of the document collection, we mine the logs that contain historical information about the users' information needs on that collection, not only can we restrict our attention to the topics of most interest to them (i.e., *hot topics*), but do a better job as well.

In fact, nowadays organizations need to know at every moment what the interests or problems of their customers are in order to respond better to their needs. In particular, by knowing which topics and corresponding documents are the ones of most interest, organizations can better organize their websites, with a hot topics service, for example, to facilitate customer searches on those greatly demanded topics. In the case of technical support centers, it has been often experienced that about 80% of customer problems refer to the same kind of problems. Therefore, by identifying the topics of these hot problems and making them readily available on the Web, customers could potentially self-solve their problems. This translates into a dramatic cost reduction in that fewer customer support experts would be needed to solve customer problems.

In this chapter, we report our approach to mine hot topics from the logs of customer support centers; in particular, we have applied this approach to the logs of Hewlett-Packard Customer Support. When a customer has a problem, he has two options to solve it. One is to simply call the customer support center and open a *case* to report the problem to a technical representative and let her figure out how to solve it. The other is to try to self-solve the problem by looking for solution documents that match the problem on the customer support website. Accordingly, there are two kinds of logs, each of which can be physically structured in different ways: a *search log* and a *case log*. The search log keeps track of the search strings that customers formulate when they try to self-solve their problems along with the documents opened after these searches, that is, the clicking paths. The case log keeps track of the cases opened by customers along with the history of actions, events, and dialogues followed while a case is open. These two logs are complementary to obtain the information about all the problems that customers have experienced. Therefore, even though they are very different from each other, we consider both in our approach. The main challenge was dealing with the various kinds of *dirtiness* or *noise* embedded in the logs in order to obtain meaningful results with the highest possible quality. Therefore, our approach to hot topics discovery is composed not only of techniques to mine search and case logs, but to deal with dirty text either by directly solving dirty features or by being robust in spite of them. The implementation of our approach is embodied in a toolbox called *HotMiner* whose general architecture is shown in Figure 6.1.

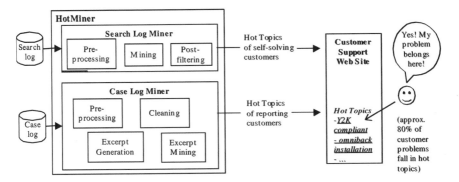

Figure 6.1. HotMiner Architecture.

Initially, we focused on the *search log* exclusively to discover hot topics of self-solving customers. Under the hypothesis that viewing the documents by the search terms used to open them it is possible to discover hot topics that match the users' perspective, we propose a novel technique based on the *search views* of documents derived from the search log. However, hot topics can only be useful if their quality is high; that is, the topics need to be clean for the users to rely on them to solve their problems. Unfortunately due to the browsing tendency of Web users, often driven by curiosity or ignorance, the clicking paths may be noisy and lead to hot topics contaminated by extraneous documents that do not belong to the topics into which their search path views push them. Our approach includes a *postfiltering* technique that considers the content of each document, *content view*, in a hot topic to identify the extraneous ones. Identifying extraneous documents is not only beneficial for obtaining higher quality topics but to pinpoint documents that are being returned as noise to some queries distracting customers from their search goal.

To complete the picture of hot topics with those of reported problems, we considered the *case log*. However, the text in case documents conformed from different fields of the case log is dirty: typos, misspellings, adhoc abbreviations, missing and ambiguous punctuation, core dumps, cryptic tables and code, in addition to being grammatically poor and nonnarrative. Hence, our approach copes with such characteristics by including techniques that directly solve some anomalies that would otherwise contaminate the results with noisy, meaningless term occurrences, and techniques that are robust in spite of the others. For example, our approach includes cleanup techniques to *normalize* typos, misspellings, and abbreviations. Terminology normalization is accomplished by using a thesaurus that is derived from the document collection by a *Thesaurus Assistant* that identifies approximate duplicates present in the documents. These approximate duplicates are synonymous words that are lexically similar. We did not consider synonyms given by lexically dissimilar words corresponding to the same concept, like those considered in Senellart and Blondel's chapter in this book, because the relevant terminology is mostly domain-specific technical terms that do not have this kind of synonyms. If this were not the case, techniques like those described in that chapter could be used to complement the thesaurus.

Furthermore, case documents are in general very long documents that report everything that happens while a case is opened, including logistic information such as "customer will report result tomorrow morning," or "cls if cu no clbk 24h" (close if customer does not call back in 24 hours). This logistic information, which is meaningless for discovering hot topics, will also introduce noise in the results. Our method includes techniques to extract relevant sentences from case documents to consolidate pure technical *excerpts* to be mined for hot topics. As a matter of fact, we first evaluated available prototype and commercial summarizers but none of them seemed to adequately deal with the characteristics of the text found in case documents, which as mentioned above is dirty and nonnarrative. Besides, these products are generic and they offer only limited capability for customizing them to specific domains, if at all, sometimes with a major effort, and still yielding results that were far from what we expected. Our excerpt generation (or summarization) method is composed of a variety of techniques capable of dealing with the characteristics of the text in case documents and making use of domain knowledge to obtain a major improvement in the quality of the excerpts obtained, that is, in the accuracy of the extracted relevant sentences. For example, the technique for sentence identification has the capability to remove code and table-formatted cryptic text that would otherwise complicate the already difficult task of identifying sentences in the presence of ambiguous and missing punctuation. In addition, since the text is grammatically poor, we are limited to using techniques that are not based on natural language. The techniques for relevant sentence identification correspond to different criteria of what makes a sentence relevant and their results can be combined in any way to obtain the final ranking of the sentences. Although the notion of *relevance* is subjective and not well understood in general, in our context it is clear that the essential requirement for a sentence to be considered relevant is to be of a technical nature and not logistic[2]. Since one of the design goals of Hotminer is to be versatile, each one of the techniques in the toolbox can be enabled or disabled. For example, techniques that make use of domain knowledge can be disabled when such knowledge is not available. The contributions of our approach are various. First, we propose a novel technique to discover hot topics from search logs. This technique combines the search view and the content view of documents to obtain high-quality hot topics that match the users' perspective. To the best of our knowledge no other technique exists that uses a *search view* of the documents to discover hot topics that, in contrast to content-based methods, match the users' perspective. It is also novel to include a postfiltering stage where the *content view* of the documents is used to boost the precision of the topics discovered from the search view of the documents. This stage corrects errors derived from noisy clicking behavior and from the approximations of the mining algorithm used to do the groupings. Second, a method to discover hot topics from case logs is proposed. The method deals with the dirtiness and mixed contents of case documents, which negatively affect the results, by including a variety of techniques like those for thesaurus generation and excerpt generation to normalize dirty variations of words and extract relevant contents, respectively. Our excerpt generation method differs from others in that it deals with dirty, nonnarrative, and grammatically poor text. It is composed of a combination of techniques to identify relevant sentences according to different criteria and without relying on grammar. The method is flexible and leaves room for customization.

Although we used several months of log content for our experiments (one month at a time), the results were not fully validated because in addition to being rather cumbersome to analyze the results obtained at different stages of the process, an ideal analysis has to be done by domain experts (whose time is precious). Therefore, we did the validation ourselves by manually inspecting a small subset of hot topics and interpreting their contents with our limited understanding of the domain. We found an average precision of 75% and although this is not perfect, the results were still satisfactory and useful to discover the hot topics of the problems that customers were experiencing, some of which did not even exist in the topic ontology of the document collection. For example, the hot topic "Y2K" did not exist as a topic in the ontology since the documents related to the Y2K problem were spread under the umbrella of different topics.

This chapter is organized in two parts, preceded by this introduction and by Section 6.2, which is a short overview of related work. The first part covers our approach to mining search logs. Section 6.3 describes the technique to discover hot topics from search strings and clickstreams and then describes the postfiltering technique to improve the quality of the hot topics. Section 6.4 shows the experimental results of applying these techniques to the search log of HP Customer

---

[2]In fact, response center engineers need a tool with the capability of extracting the technical content of case documents to assist them in the creation of certified documents.

Support. The second part covers our approach to mining case logs and follows the same structure as the first part, that is, first a technical description and then experimental results. Section 6.5 describes the method for obtaining excerpts of relevant sentences from cases. Section 6.6 gives a flavor of the results obtained by applying the sentence extractor to a case document. Section 6.7 briefly describes how hot topics are discovered from case excerpts. We finalize by presenting our conclusions in Section 6.8.

6.2   Related Work

From the Introduction it can be surmised that to discover hot topics from search and case logs we need to deal with different aspects of the problem. Therefore, our approach is composed of a variety of techniques that separately tackle these aspects. In this section, we briefly review work related to each of our techniques.

Document categorization can be done manually or automatically. In manual categorization, experts identify the topic of each document in a collection. The resulting topic hierarchies/ontologies are intuitive and the users navigate through them naturally and efficiently, as in Yahoo!. However, categorizing documents manually is highly time consuming and costly, so emphasis has been put on finding ways of doing it automatically. There are two kinds of automatic categorization. In the supervised categorization, called *classification*, the categories (topics) are predefined and a subset of category-labeled documents is used to train a classification algorithm to learn how to classify documents into these categories. Some work in this direction is described in [DPHS98, LSCP96, YY99]. Since we do not know the categories (hot topics), our work does not fall in this category. The other kind of automatic categorization is the unsupervised one, where not only documents need to be categorized, but the categories themselves need to be discovered. In this case, a *clustering* algorithm [DHS01] is typically used to discover the categories implicit in the collection, as originally proposed in [Sal89]. Numerous techniques such as those described in [CKPT92, Wil88, SKK00] have followed the clustering approach but the results have never been comparable to the ones obtained manually; that is, the categories are often unnatural and not intuitive to the users. In contrast, we focus on discovering hot categories from a novel view of documents based on the search terms used to open them (search view) or from excerpts comprised of the most representative sentences of case documents.

Work on *automatic correction of words* in text is surveyed in [Kuk92]. Some methods for identifying misspelled words in text require words to be tagged by a part-of-speech tagger [Too00] and [EE98]. These methods are not applicable to our work since we are dealing with dirty text, which yields poor results when tagged by a part-of-speech tagger.

There exist numerous spelling checkers for the most popular operating systems. We chose to use the UNIX built-in spelling checker "spell" [McI82] to identify misspellings. Spelling checkers such as "Ispell" [Kue87] also make suggestions for misspelled words but we chose instead to use the Smith–Waterman algorithm [SW81] for finding the "approximate duplicate" relationship of misspellings to words in the vocabulary of the collection because of the robustness of the algorithm and the proprietary-specific terminology of our domain.

The technique used for the *removal of table-formatted text* is based on work reported in [Sti00] on the identification of tables in PDF documents. Very little has been done in the area of *removing code fragments* from text documents, barring the removal of dead code from computer programs.

Not too much work on *sentence boundary detection* has been reported. Most of the techniques developed so far follow either of two approaches: the rule-based approach which requires domain-specific handcrafted lexically based rules [Nun90], and the corpus-based approach where models are trained for identifying sentence boundaries [RR94], sometimes including part-of-speech tagging in the model [SSMB99]. Since we do not have an annotated corpus, we manually handcrafted some heuristic rules that we identified by inspection.

Research on *summarization* has been done since the late 1950s [Luh58], but it was only in the 1990s that there was a renaissance of the field. A compendium of some of the most representative approaches is found in [MM99]. As stated there, according to the level of processing, summarization techniques can be characterized as approaching the problem at the surface, entity, or discourse levels. Surface-level approaches represent information in terms of shallow features. Entity-level approaches tend to represent patterns of connectivity of text by modeling text entities and their relationships, such as those based on syntactic analysis [CHJ61] or on the use of scripts [Leh82]. Discourse-based approaches model the global structure of the text and its relation to communicative goals, such as [Mar97] who uses the rhetorical structure in the discourse tree of the text, and [BE99] who uses WordNet to compute lexical chains from which a model of the topic progression is derived. Most of the entity- and discourse-based approaches are not applicable in a domain such as ours where the specific lingo, bad use of grammar, and nonnarrative nature of the text make it infeasible to apply existing lexical resources such as WordNet, scripts of stereotypical narrative situations, or in-depth linguistic analysis. The techniques used in our method are surface- and entity-level techniques.

Part I: Mining Search Logs

As stated in the Introduction, we initially focused on discovering hot topics of the customers who try to self-solve their problems. Data related to the search performed by customers to find the documents with the solutions to their problems are captured in search logs. In this part of the chapter we describe techniques for mining these logs to discover hot topics that match the customers' perspective.

6.3   Technical Description

The main idea is to use a view of the documents derived from the search terms used to open them, called *search view*, to discover the hot topics of customers' problems and associated documents. Then, to boost the precision of these topics, we propose to use the traditional view of the documents derived from their contents to identify those documents that fall into the wrong topics and filter them out. The method is comprised of four main stages: preprocessing, clustering, postfiltering, and labeling, as shown in Figure 6.2.

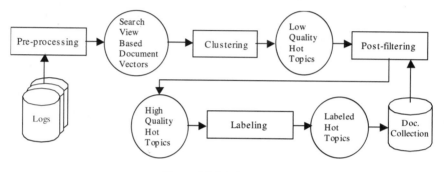

Figure 6.2. Methodology.

6.3.1   Preprocessing

In this stage, the raw material of the search log is processed to extract the relevant data to derive the search views of the documents that are subsequently converted into a form suitable for data mining.

Data Collection

The log provides complete information about the searches performed by users: the queries formulated by them, as well as the documents that they opened after the queries. This data collection step can take different forms depending on the physical configuration of the search logs. HP Electronic Customer Support, which is the site HP customers used to visit to self-solve their problems, has a search log composed of three separate logs: a usage log, a query log, and a web log. These logs are described next.

A *usage log* records information on the actions performed by the users, for example, doing a search (doSearch) or opening a document from the list of search results (reviewResultsDoc). A usage log entry looks like the following.

Timestamp	Access Method	User Id	Remote Machine	Subject Area	Transaction Name
011000 08:35:14	HTTP	CA1234	123.28.53.1	atq	doSearch
011000 08:36:55	HTTP	CA1234	123.28.53.1	atq	reviewResultsDoc

A *query log* records information on each query posed after a doSearch action. A search log entry might look like the following.

Timestamp	User Id	Remote Machine	Mode	Query String
011000 08:35:39	CA1234	123.28.53.1	Boolean	y2k

A *web log* records information on each web page (i.e., document) opened. An entry might look like the following.

Timestamp	Remote Machine	URL
011000 08:43:37	123.28.53.1	"GET /atq/bin/doc.pl/?**DID=15321** HTTP/ 1.0"

It is necessary to correlate the usage, search, and web logs to obtain a relation between the searches performed by the users and the corresponding documents opened as a result of the searches. First, the usage and search logs are correlated to obtain the time at which a query was formulated. Then, correlation with the web log takes place to look for all the documents that were opened after the subsequent reviewResultsDoc action in the usage log but before the timestamp of the next action performed by that user. Sequential numbering had to be introduced to handle the lack of synchronization of the clocks in the different logs. This step results in a document-query relation that looks like the following.

Document	Query String
15321	Y2k; 2000 compatibility; year 2K ready;
539510	Omniback backup; recovery; OB;
964974	sendmail host is unknown; sender domain sendmail; mail debug; sendmail; mail; mail aliases;

Since we are interested in obtaining topics that match the users' perspective, instead of viewing documents by their contents, which is the usual way, we propose to view them by the collection of queries that have been posed against them; that is, for each document we associate all the queries that resulted in opening that document. This is called a *search view* or *query view* of a document. We can restrict the document-query relation to a given number of the hottest (i.e., more frequently accessed) documents. This is useful when the hot topics service only contains hot documents in the hot topics.

Data Cleaning

The entries of the document-query relation need to be scrubbed, which might involve removing some entries from the relation, normalizing query terms, removing special symbols, eliminating stop-words, and stemming (see Section 6.3.3 – Data Cleaning).

	Word 1	Word 2	...	Word n
Doc 1	w_{11}	w_{12}	...	w_{1n}
Doc 2	w_{21}	w_{22}	...	w_{2n}
...	...	...	...	...
Doc m	w_{m1}	w_{m2}	...	w_{mn}

Table 6.1. *V* Vectors.

Data Transformation

Each cleaned entry of the document-query relation has to be transformed into a numerical vector for it to be mined in the second stage. From this transformation we obtain a set of vectors V, with one vector per document consisting of elements w_{ij}, which represent the weight or importance that word j has in document i. This weight is given by the probability of word j in the set of queries posed against document i as shown in the following example.

Set of queries for document 964974:
{sendmail, mail, sender domain sendmail, mail aliases, sendmail host unknown, mail debug}
Vector representation for document 964974:

sendmail	mail	sender	domain	aliases	host	unknown	debug
0.25	0.16	0.08	0.08	0.08	0.08	0.08	0.08

Since the number of different words that appear in a large collection of queries can be quite large, a *feature selection* technique is used to select a set of the most relevant words to be used as the dimensions of the vectors. We are interested only in hot topics, therefore we use the simple technique of selecting as features the top n query words that appear in the search view of at least two documents. Phrases with more than one word could be used as well, but our experiments indicated that they did not add any value.

The set of vectors V is organized into a matrix where each row corresponds to a document vector and each column corresponds to a feature as shown in Table 6.1.

6.3.2   Clustering

In this stage, the vector representation of the search view of the documents is mined to discover hot topics that match the user's perspective. Groups of documents that are similar with respect to the users' information needs will be obtained by clustering [DHS01] the documents representing a collection of points S in a multidimensional space R^d where the dimensions are given by the features selected in Step 6.3.1 – Data Transformation. Numerous definitions of clustering exist and for each definition there are multiple algorithms. To illustrate, let us take a definition known as the *k-median* measure: given a collection of points S in R^d

and a number k of clusters, find k centers in R^d such that the average distance from a point in S to its nearest center is minimized. While finding k centers is NP-hard, numerous algorithms exist that find approximate solutions. Any of these algorithms, such as k-means, farthest point, or SOM maps can be applied to the *Documents X Words* matrix given in Table 6.1. The interpretation of a cluster so produced is that two documents are in the same cluster if they correspond to similar searches; that is, customers opened them after similar queries.

Although our method is independent of the clustering definition and algorithm used, experimentation is recommended to find the algorithm that works best for the domain at hand, that is, finds the most accurate hot topics.

6.3.3 Postfiltering

The *search view* of the documents relies on the assumption that users open documents that correspond to their search needs. However, in practice users often exhibit a somewhat arbitrary or even chaotic clicking behavior driven by curiosity or ignorance. Under this behavior they open documents that do not correspond to their search needs but that the search engine returned as a search result due to the presence of a search term somewhere in the document. In this case, the document's search view erroneously contains the search string and therefore it will be associated with documents whose search view also contains that string, probably ending up misplaced in the wrong cluster. For example, a document on the product Satan, which is software for security, is returned in the list of results for the query strings "year 2000" or "year 2000 compliance" since it contains a telephone extension number "2000," although it is not related to "year 2000." Users tend to open the Satan document probably misled by the name Satan which they might interpret as something catastrophic related to Y2K compliance or just driven by curiosity as to what it could be about. Thus the search view of the Satan document contains the search string "year 2000" which makes it similar to documents related to Y2K and whose search view also contains the string "year 2000." Hence, the Satan document ends up in a cluster corresponding to the Y2K topic. These *extraneous* documents that appear in topics to which they do not strictly belong constitute noise that has a negative impact on the quality of the topics. Since users will not rely on a hot topics service with low quality, it is necessary to clean up the hot topics by identifying extraneous documents and filtering them out. This is accomplished by computing the similarity of the documents according to their content (i.e., their *content view*), and designating as candidates for filtering those with a low similarity to the rest of the documents in their cluster. The steps comprising this stage are given next.

Data Collection

In this step the actual contents of the documents in the clusters are retrieved. As mentioned above, the postfiltering stage builds upon the views obtained from the text conforming these documents.

Data Cleaning

The contents of the documents are scrubbed as in Section 6.3.1 – Data Cleaning, but here header elimination and HTML stripping are necessary, followed by special symbol removal, stop-word elimination, normalization, and stemming.

- Header elimination: removes metainformation added to the documents that is not part of the content of the documents themselves.

- HTML stripping: eliminates HTML tags from the documents.

- Special symbol removal: removes symbols such as *, #, &, and so on.

- Stop-word elimination: eliminates words such as "the", "from", "be", "product", "manual", "user", and the like that are very common and therefore do not bear any content. For this we use both a standard stop-word list available to the IR community and a domain-specific stop-word list that we created by analyzing the contents of the collection.

- Normalization: attempts to obtain a uniform representation of concepts through string substitution, for example, by substituting the string "OV" for "OpenView".

- Stemming: identifies the root form of words by removing suffixes, as in "recovering" which is stemmed as "recover".

Data Transformation

The contents of the documents are transformed into numerical vectors similar to those in Section 6.3.1 – Data Transformation. However, here the set of different terms is much larger and the technique that we use for selecting features consists of first computing the probability distributions of terms in the documents and then selecting the l terms in each document that have the highest probabilities and that appear in at least two documents. Also, the weight w_{ij} for a term j in a vector i is computed differently, by using our slight modification of the standard TF-IDF (Term Frequency – Inverse Document Frequency) measure [Sal89] given by the formula:

$$w_{ij} = k_{ij} \left[tf_{ij} \log(n/df_j) \right], \tag{6.1}$$

where n is the number of documents in the collection, tf_{ij} corresponds to the frequency of term j in document i, df_j is the document frequency of term j, and K is a factor that provides the flexibility to augment the weight when term j appears in an important part of document i, for example, in the title of the document. The default value for k is 1. The interpretation of the TF-IDF measure is that the more frequently a term appears in a document, the more important it is, but the more common it is among the different documents, the less important it is since it loses discriminating power.

The weighted document vectors form a matrix like the one in Table 6.1, but the features in this case are content words as opposed to query words.

Extraneous Document Identification

This step is composed of two substeps: similarity computation and filtering.

Similarity Computation

The similarity of each document d_i to every other document d_j in the cluster where it appears is computed. The cosine distance is used to compute the cosine angle between the vectors in the multidimensional space R^d as a metric of their similarity

$$\cos(d_i, d_j) = (d_i \times d_j)/\sqrt{d_i^2 \times d_j^2} \, .$$

From these individual similarities, the average similarity AVG_SIM (d_i) of each document with respect to the rest of the documents in the cluster is computed. Let n be the number of documents in a cluster; then

$$\text{AVG\_SIM}(d_i) = (1/n) \sum_{j=1, \, j \neq i}^{n} \cos(d_i, d_j) \, .$$

A cluster average similarity AVG_SIM (C_k) is computed from the average similarities of each document in the cluster C_k;

$$\text{AVG\_SIM}(C_k) = (1/n) \sum_{i=1}^{n} \text{AVG\_SIM}(d_i) \, .$$

Filtering

The document average similarity AVG_SIM (d_i) is used as a metric of how much cohesion exists between a given document d_i and the rest of the documents in the cluster. If the similarity is high it means that the document fits nicely in the cluster where it appears, but if the similarity is low it means that it is an extraneous document in the cluster where it was erroneously placed. Now the question is how "low" the similarity has to be for a document to be considered extraneous.

We need a mechanism for setting a similarity threshold that establishes the boundary between extraneous and nonextraneous documents in a cluster. This threshold will depend on the magnitude of the deviation from the cluster average similarity AVG_SIM (C_k) that will be allowed for the average similarity AVG_SIM (d_i) of a document D_i. For this we compute the standard deviation for a cluster C_k:

$$S(C_k) = \left(\frac{1}{n-1} \sum_{i=1}^{n} [\text{AVG\_SIM}(d_i) - \text{AVG\_SIM}(C_k)]^2 \right)^{1/2} \, .$$

Then, we standardize each document average similarity AVG_SIM (d_i) to a Z-score value

$$Z(\text{AVG\_SIM}(d_i)) = [\text{AVG\_SIM}(d_i) - \text{AVG\_SIM}(C_k)] / S(C_k)$$

to indicate how many standard deviations the document average similarity AVG_SIM (d_i) is off the cluster average similarity AVG_SIM (C_k).

By inspecting the Z-score values of known extraneous documents we can set the threshold for the Z-score value below which a document is considered a candidate extraneous document in the cluster where it appears. To support the decision made on the threshold setting we can use the Chebyshev theorem, which says that for any data set D and a constant $k > 1$, at least $1 - (1/k^2)$ of the data items (documents in our case) in D are within k standard deviations from the mean or average. For example, for $k = 1.01$ at least 2% of the data items are within $k = 1.01$ standard deviations from the mean, whereas for $k = 2$ the percentage increases to 75%. Therefore, the larger k is, the more data items are within K standard deviations and the fewer documents are identified as extraneous candidates. The opposite holds as k gets smaller because fewer documents are within a smaller number of standard deviations and only those exceeding this number are considered extraneous candidates. This translates into the typical trade-off between precision and recall. If we want more precision (i.e., the proportion of true extraneous candidates to the total number of candidates), the price paid is on the true extraneous documents that will be missed (false negatives) which decreases the recall. On the other hand, if we want to augment the recall (i.e., the proportion of true extraneous candidates to the total number of true extraneous documents), precision will decrease because documents that are not extraneous are identified as extraneous candidates (false positives).

Another way to assess the effect of the threshold value setting is to use the Coefficient of Dispersion of a cluster

$$V(C_k) = [S(C_k)/AVG\_SIM(C_k)] \times 100 ,$$

which expresses the standard deviation as a percentage. It can be used to contrast the dispersion value of a cluster with and without the extraneous candidates.

The extraneous candidates can be filtered either manually or automatically. Automatically means that all the extraneous candidates will be eliminated from their clusters. In this case, the Z-score threshold setting is more critical because the false extraneous candidates will be eliminated while at the same time true extraneous documents not identified as candidates will be missed. The manual mode is meant to assist a domain expert for whom it is easier to pinpoint false extraneous candidates than to identify missed extraneous ones. Therefore, in this mode we set the threshold for high recall, and have the domain expert eliminate the false candidates to raise the precision. The candidates could be highlighted when the documents of a hot topic are displayed, so that the domain expert could easily visualize the candidates and either confirm or reject them as extraneous to the topic.

6.3.4   Labeling

Once the clusters have been cleaned by removing extraneous documents, a label has to be given to each cluster to name the topic that it represents. Labels can be assigned either manually or automatically. Manual labeling is done by a human who inspects the documents in a cluster and, using his experience and background knowledge, assigns a semantic label to the topic represented by the cluster. Automatic labeling is done by analyzing the words in the documents of a cluster to find the L most common words that appear in at least half of them and using those L words to label the cluster. Those clusters for which there is no such label could be simply discarded. If there is more than one cluster with the same label they are merged and postfiltering is applied to the resulting cluster. In the experiments section we describe different ways of doing the labeling according to the kind of words and parts of the documents that are considered.

6.4   Experimental Results

The method used to mine hot topics described in the previous section has been implemented as part of a toolbox named HotMiner. In this section we show some results obtained by applying HotMiner to discover hot topics from the search log of HP's Electronic Customer Support website. This site makes a collection of one million documents available to HP's registered customers. When a customer has a software problem, she can either submit a case to have a support engineer find a solution for her problem, or she can try to self-solve it by searching the collection with the hope of finding a solution document for the problem. If, according to the customer's experience, self-solving can be done efficiently, it is likely that she will return to the site to self-solve other problems when they arise. However, if she finds it difficult to find the right solution document, most likely she will quit the site, open a case, and not attempt to self-solve her problems ever again.

The search process required for self-solving problems can be done either by navigating through a topic ontology when it exists, or by doing a keyword search. The disadvantage of using a topic ontology is that it frequently does not match the customer's perspective of problem topics, making the navigation counterintuitive. This search mode is more suitable for domain experts, such as support engineers, who have a deep knowledge of the contents of the collection of documents and therefore of the topics derived from it. In the keyword search mode the user knows the term(s) that characterize his information needs and he uses them to formulate a query. The disadvantage of this search mode is that the keywords in the query not only occur in documents that respond to the user's information need, but also in documents that are not related to it at all. Therefore, the list of search results is long and tedious to review. Furthermore, the user often uses keywords that are different from the ones in the documents, so he has to go into a loop where queries have to be reformulated over and over until he finds the right keywords. However, by applying HotMiner to the various ESC logs, we discover hot topics that match the user's perspective and that group documents corresponding to the hottest problems that customers are experiencing. When these hot topics are made directly available on the website, customers can find solution documents for the hottest problems straightforwardly. Self-solving becomes very efficient in this case.

Taking November of 1999 as a representative month, the number of documents from the Electronic Customer Support website that HP customers opened was 49,652, of which 12,678 were distinct, and the number of queries was 27,884. According to support engineers, 80% of the customer calls correspond to 20% of the problems, so they needed a way to identify these hot problems and publish them on the website along with the appropriate solution documents. HotMiner is the tool that we built to assist them in discovering this information.

HotMiner implements each of the steps according to the description given in Section 6.3. However, some steps for which alternatives exist and that were left as free choice in that section are further described below. Some of these steps required experimentation to evaluate the alternatives.

- The normalization step was limited to a few words that were manually normalized because the Thesaurus Assistant and Normalizer, described in Section 6.5, that help to identify and normalize variations of words did not exist at the time our experiments were conducted.

- In the clustering step and for the experiments on ESC log data we used the SOM_PAK [KHKL96] implementation of Kohonen's Self-Organizing Map (SOM) algorithm [Koh92]. The output of SOM_PAK allows for an easy implementation of the visualization of the map, which in turn facilitates the analysis of the evolution of topics. Our visualization implementation produces a map that looks like the partial map shown in Figure 6.3. In this map each cell represents a grouping of related documents; the more related the groupings are, the closer their corresponding cells appear in the map. As can be observed in Figure 6.3, there are several cells that have overlapping labels, as in the five left upper cells whose labels contain the word "sendmail". Overlapping labels gave us an indication that the cells represent groupings that might constitute a topic. In order to discover these topics, we applied SOM again, but this time on the centroids (i.e., representatives) of the groupings obtained in the first SOM application. We expected that the centroids would cluster naturally into higher-level topics that generalize the first-level topics (subtopics). However, in some cases the resulting groupings did not make much sense (i.e., subtopics that were unrelated were grouped together). One possible reason is that the relationships between the centroids found by SOM are not intuitive from the problem-solving perspective, and another possible reason is the diverse uses of the same term. To visualize these higher-level groupings, we labeled the cells in different colors according to the higher-level grouping to which they belong as identified by SOM. Since this book is in black and white, in Figure 6.3 we show these higher-level groupings by drawing an enclosing line around the cells belonging to each high-level grouping.

- In the postfiltering stage, the vector representations of the contents of the documents can be derived in different ways depending on which words are taken into account for their content view and the value of k in the modified

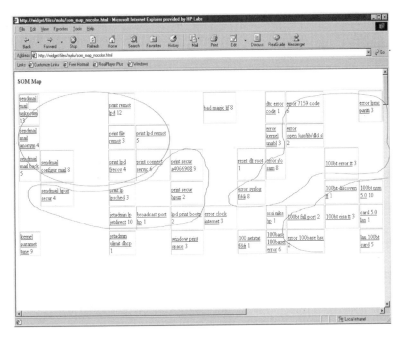

Figure 6.3. Partial cell map.

TF-IDF formula used to transform these views to vectors. Alternatives that were experimented with are:

(a) occurrences of non-stop-words in the titles of the documents;

(b) occurrences of non-stop-words in the entire contents of the documents;

(c) same as (b) but using $k > 1$ in Eq. (6.1) when the word appears in the title;

(d) query words (of the grouping to which the document belongs) that appear in the document; and

(e) occurrences of the first p words in the documents (solution documents usually contain the problem description at the beginning).

With each of these representations, the similarity values described in *Similarity Computation*, Section 6.3.3 – Extraneous Document Identification, were computed and compared for some clusters. The similarity values obtained in alternatives (a), (c), and (e) were the least accurate. In particular, in alternative (c) we observed that the larger the value of K is, the higher the similarity values for extraneous documents are, which is wrong. The alternative that gave the highest accuracy for similarity values was (b).

- The labeling step also offers several alternatives. One is to derive the labels from the *query* words in the vector representation of the query view of the documents. The other is to use the *content* words in the vector representation of the content view of the documents. In the second case (i.e., using content words), there are two independent sets of additional options that can be combined in any way. The first set of options gives a choice for the kind of content words to use for labeling: non-stop-words, nouns, or noun phrases. For nouns and noun phrases we require a PoS (part-of-speech) tagger and a noun phrase recognizer. We used the ones provided in the LinguistX platform library from Inxight [Inx]. The second set of options gives us a choice to use only the titles of the documents or the whole contents of the documents to extract from them the non-stop-words, the nouns, or the noun phrases.

Our experiments show that by using all non-stop-words in the document titles we obtain labels that better describe the topic represented by a cluster. This is not strange since titles convey the most important information about the contents of the documents and therefore constitute a first-level summary of the documents. However, not all the documents in the collection have meaningful titles and the topics where this happens often have meaningless labels. According to our experiments, to avoid this situation while at the same time obtaining "good" labels, the labeling technique should use all words that are not stop-words that appear in the whole contents of the documents. It seems counterintuitive that using all non-stop-words performed better than using noun phrases, but as the following example from our experiments shows, noun phrases are more restrictive. For a topic on "shared memory" problems the label "memory" was assigned when nouns or noun phrases were used, but "shared" and "memory" were the labels when non-stop-words were used. This is because although the documents are related to shared memory, most of the titles of the documents in that cluster contain the word "memory", others contain the word "shared", and only a few had the noun phrase "shared memory". When we use nouns or noun phrases, "memory" is found to be the most common noun and noun phrase and therefore it is the one used for labeling; "shared" is neither a noun nor a noun phrase so it does not even count. When we use all non-stop-words the most common words are "memory" and "shared", which are the ones used for the label when the number l of most common words to be used for a label is set to two. However, labeling is a difficult problem and our results often did not give us completely meaningful labels. For example, in Figure 6.4 the label "y2k 10.20 patch" might be interpreted as a topic on patches for HP-UX 10.20 for Y2K compliance, which seems fine according to the documents in that topic. However, in Figure 6.5 the label "sendmail mail unknown" is not as easy to interpret and can only be interpreted as a topic related to "sendmail" problems.

Once the topics are found and labeled, the corresponding SOM map visualization is created on a web page that lets us analyze the topics obtained (this visualization is not intended for the customers). By clicking on the label of a cell, the contents of the topic in terms of its hottest documents is displayed in another web page as shown in Figures 6.4 through 6.6. In such a page all the documents that fall into the corresponding cluster are listed, and the ones that the postfiltering stage

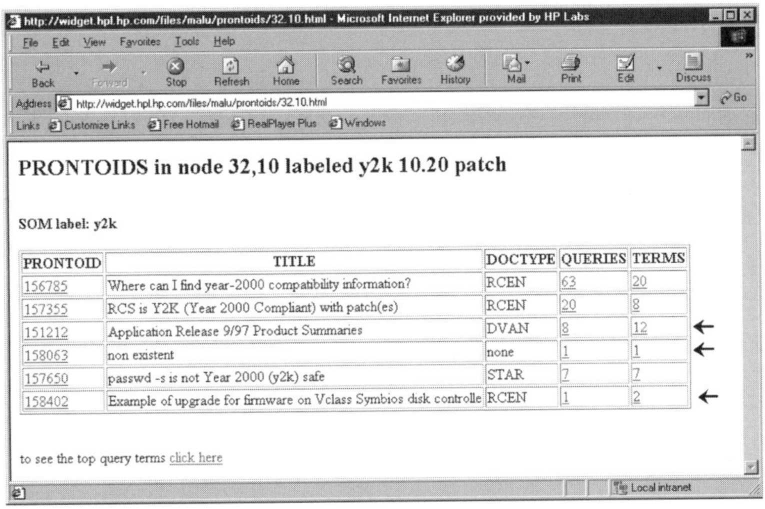

Figure 6.4. Successful postfiltering of extraneous documents.

found as extraneous appear in red.[3] Those three figures also show that postfiltering can never be perfect, and that a compromise has to be made between recall and precision. Figure 6.4 is an example where postfiltering was totally successful in identifying all the extraneous documents (documents 151212, 158063, and 158402). Figure 6.5 shows partial success in identifying extraneous documents since not only were all the extraneous documents identified (document 150980), but nonextraneous documents were identified as extraneous as well (documents 156612 and 154465). Finally, Figure 6.6 is an example of a total failure in identifying extraneous documents since there are no extraneous documents but two of them were identified as such (documents 159349 and 158538). This might happen in any cluster that does not contain extraneous documents since the average similarity of some documents, even if it is high, may be under the similarity threshold when the standard deviation is very small. To avoid this problem, the standard deviation in each cell has to be analyzed and if it is very small postfiltering will not be applied to the cell.

In support of our hypothesis that the topics discovered with this method match the users' perspective, we make the following observation: several topics obtained did not appear in the existing topic hierarchy, either because they were just emerging, like the topic Y2K, or because the support engineers need a different perspective of the content of the collection that corresponds to organizational needs, or simply because they were not aware of the way users perceive some problems. In any case, it was interesting to see which topics were hot at a given moment, how much

---

[3]Since the printing of this book does not allow colored figures, candidate extraneous documents are shown pointed to by a small arrow.

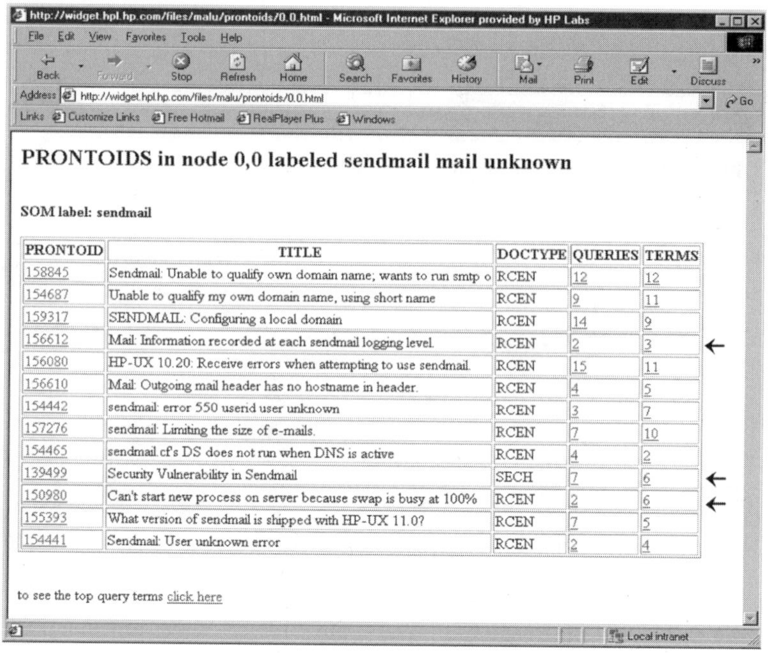
PRONTOIDS in node 0,0 labeled sendmail mail unknown

SOM label: sendmail

PRONTOID	TITLE	DOCTYPE	QUERIES	TERMS
158845	Sendmail: Unable to qualify own domain name; wants to run smtp o	RCEN	12	12
154687	Unable to qualify my own domain name, using short name	RCEN	9	11
159317	SENDMAIL: Configuring a local domain	RCEN	14	9
156612	Mail: Information recorded at each sendmail logging level.	RCEN	2	3
156080	HP-UX 10.20: Receive errors when attempting to use sendmail.	RCEN	15	11
156610	Mail: Outgoing mail header has no hostname in header.	RCEN	4	5
154442	sendmail: error 550 userid user unknown	RCEN	3	7
157276	sendmail: Limiting the size of e-mails.	RCEN	7	10
154465	sendmail cf's DS does not run when DNS is active	RCEN	4	2
139499	Security Vulnerability in Sendmail	SECH	7	6
150980	Can't start new process on server because swap is busy at 100%	RCEN	2	6
155393	What version of sendmail is shipped with HP-UX 11.0?	RCEN	7	5
154441	Sendmail: User unknown error	RCEN	2	4

to see the top query terms click here

Figure 6.5. Partial success in identifying extraneous documents.

corresponding content existed, and the clicking behavior of customers when trying to self-solve problems belonging to a given topic.

To fully evaluate the results we would need to analyze the content of each hot topic and its documents, which would require a prohibitive amount of effort. As an alternative, we did a limited evaluation of the results. We randomly sampled 10% of the hot topics and found that about 80% of the sample topics made sense. The relationship of the documents in the remaining 20% was not apparent to us due to our lack of domain knowledge or because although their search views were related, their content views were not. In any case, if automatic elimination of topics is enabled, most of the meaningless hot topics could be automatically eliminated at two different times: in the postfiltering step by setting thresholds for cluster average similarity, AVG_SIM (C_k), and cluster standard deviation, S (C_k), to values that experimentally prove to get rid of most noncohesive topics, and in the labeling step when no meaningful label is found for a topic. Nevertheless, a human should be involved in the loop to inspect the final hot topics before making them public.

Part II. Mining Case Logs

As mentioned in the Introduction, customers do not always try to self-solve their computer-related problems on the customer support website. They often open a case via a phone call, e-mail, or the Web. In fact, even when they try to self-solve their problems, often they are not successful in their search for the right

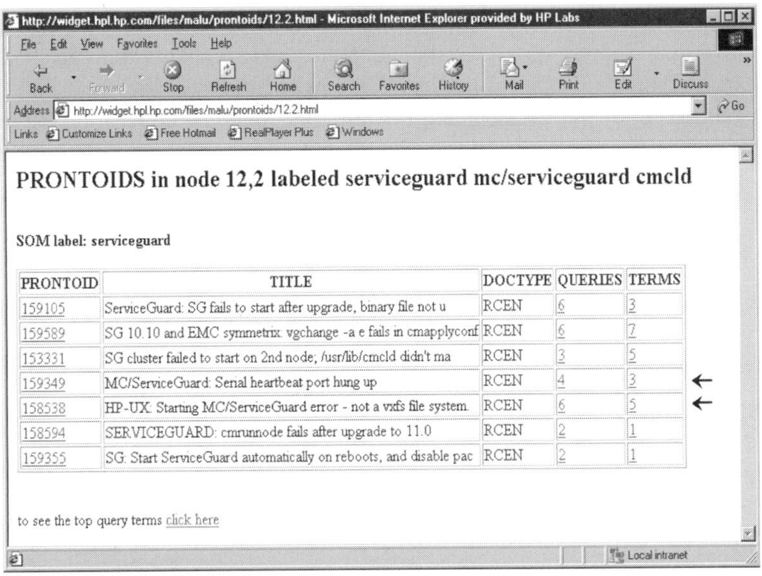

Figure 6.6. Failure to identify extraneous documents.

documents or they simply get tired of searching and end up opening a case. While a case is open, the problem and anything related to finding a solution to it are captured in a case log. Therefore, to get a complete picture of the hot topics of customers' problems, it is not enough to mine searches and clickstreams in search logs, but also problem reports or case documents (called *cases* from here on) in case logs. Cases are unstructured, informal, usually long documents written quickly and containing anomalous or dirty text: typos, misspellings, adhoc abbreviations, code, core dumps, cryptic tables, ambiguous and missing punctuation, and bad use of English grammar. They contain not only technical facts related to the problem but also the logistics followed during the solving process. These characteristics of cases in their original form make them inadequate to be mined for hot topics. Therefore, in our approach we first clean the cases from as many anomalies as possible and then extract the most relevant sentences to generate short cleaner excerpts that are more suitable for mining hot topics.

The main focus in this part of the chapter is on the cleaning and generation of excerpts. Mining the excerpts for hot topics is only briefly described at the end since the process basically reuses steps described in the first part.

6.5   Technical Description

Our approach to generating excerpts from cases is an instantiation of our methodology to mine dirty text in three stages: application-independent cleaning, instantiated by a *Thesaurus Assistant*; application-driven cleaning, instantiated by

Code and *Table Removers* of the *Sentence Identifier*; and content extraction, instantiated in a *Sentence Extractor* that generates excerpts. This approach has been implemented as the FlexSum[4] toolbox that can be integrated into HotMiner, or as a standalone set of independent and flexible tools that can be customized for the application and domain at hand. Its general architecture is shown in Figure 6.7.

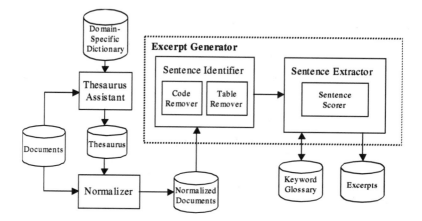

Figure 6.7. FlexSum architecture.

To deal with "dirty" lexical variations of words (i.e., typos and abbreviations), and account for them as occurrences of the same word, a thesaurus is needed. However, since relevant terms in our domain are so specific, existing thesauri are not useful. We need to generate a thesaurus that identifies correspondences of these variations to specific technical terms and to Standard English as well, and that assists a domain expert in the selection of the normalized (i.e., clean) form of dirty variations of words. This cleaning stage is *application independent* and constitutes the first stage of the methodology.

The second stage is again cleaning, but this time the cleaning is *application* and *domain driven*. For the summarization application and the *case* domain, code and cryptic tables are removed prior to sentence identification. This dirty text complicates even more the already difficult task of identifying sentences when the text has missing and ambiguous punctuation marks. In addition, we found that in the *case* domain relevant information for summarization is rarely found in code or table form, so we do not even consider their content at all. For domains where this is not the situation, code and tables may be removed only for sentence identification purposes, leaving them otherwise available for further analysis.

Once the cleanup task has been performed, the third stage of the methodology, *content extraction*, proceeds by analyzing and identifying sentences under various

---

[4]The name FlexSum is intended to convey the idea of a flexible summarization, that is, a summarizer flexible enough to be configured to any particular kind of document (i.e., clean/dirty, narrative/nonnarrative, generic/specific, and so on).

criteria, corresponding to different techniques, to assess their relevance. Then the most relevant sentences are extracted to conform the excerpts. Figure 6.8 shows an example of the dirty features of a *case*.

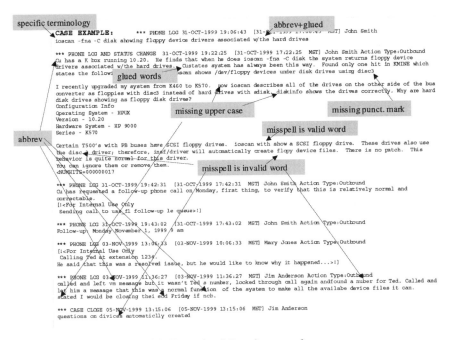

Figure 6.8. Example of dirty features of a *case*.

6.5.1   Thesaurus Assistant

Normalization of a document to remove typos, misspellings, and abbreviations requires the existence of a *thesaurus* that can be used to replace words that are misspelled, have typographical errors, or are nonstandard abbreviations (*dirty words*), with correctly spelled words or standard abbreviations (*reference words*). Creation of the thesaurus involves creating a *reference word list* as well as a *dirty word list* and identifying which of the words in the dirty word list are *approximate duplicates* of words in the reference list. Those dirty words that are within a specified edit distance from a word on the reference list are considered to be approximate duplicates of that reference word.

Having identified the approximate duplicates, the next step is to determine which of the approximate duplicates of the reference terms are to be included in the thesaurus as *synonyms* of the reference term (this step requires a domain expert and cannot be fully automated). It is these *synonyms* that are replaced by the reference term when the documents are normalized.

Creation of the Reference Word List and the Dirty Word List

To create the reference word list, each document in the corpus is broken down into its individual words. In our method, for a word to be on the reference list, it must either pass a spell-checker, or be in a *domain-specific term dictionary* [5] since valid technical terms typically do not appear in an ordinary spell-checker's dictionary. If neither of these conditions is met, the word is considered to be a dirty word.[6]

Finding Approximate Duplicates

Under the assumption that words which are misspelled or contain typographical errors only differ from the correct word by a few keystrokes, and that abbreviations only differ from their complete counterparts by some missing keystrokes, we use the edit distance between two words as a measure of whether they are approximate duplicates of each other. The particular algorithm we use to calculate the edit distance is the Smith–Waterman algorithm [SW81]. This algorithm determines the edit distance by the minimum number of edit operations (insertion, deletion, substitution, and permutation) required to transform one word into another. The edit distance is calculated as a number between zero and one, where an edit distance close to zero means the two words are far apart and an edit distance close to one means the two words are very close to each other. By setting a *threshold* on the edit distance, those words that are far apart can be filtered out.

This edit distance threshold is experimentally determined by creating a benchmark and running the Smith–Waterman algorithm on it with varying threshold values until one is found that includes a significant number of words which are truly synonyms while rejecting words that are close in edit distance but are actually not valid synonyms. The benchmark consists of a list of reference terms and, for each reference term, a number of candidate terms that are either close in edit distance to the reference term or bear some resemblance to the reference term. Half of the candidate terms are *valid* synonyms of the reference term and half are not. Ideally, Smith–Waterman should identify exactly those valid synonyms as approximate duplicates, and none of those that are not. In our experiments, the threshold of 0.67 was found to maximize recall while offering adequate precision.

It should be pointed out that for short words of four characters or less the threshold value of 0.67 breaks down; that is, too many words that are not approximate duplicates of the short word are found. It is for this reason that the edit distance threshold really needs to be a function of word length. Intuitively, the threshold should approach a value of one as the word length decreases. We have not determined a satisfactory function for the threshold value as yet.

---

[5] If a domain-specific term dictionary does not exist or is incomplete, it can be built or improved upon by manually examining the list of words that are of highest frequency and yet do not pass the spell-checker.

[6] For the sake of clarity we refer to words but in fact, it also applies to collocations of any length where at least one of its component words is dirty. We used lengths equal to one and two because the technical terms of our domain are of length two at most.

To find the approximate duplicates to the words on the reference list, each word on the dirty word list is paired in turn with each word on the reference word list and the edit distance is calculated. If the edit distance exceeds the threshold set above, the word from the dirty word list is added to the list of approximate duplicates for the word from the reference list. The final output is a list of reference words along with their corresponding lists of approximate duplicates, as shown below for the reference term "service guard":

service guard   serviceguard, servicegaurd, service gaurd, mc/serviceguard, service gard, servicegard

Thesaurus Creation

The output of the approximate duplicate computation needs to be manually filtered to select valid entries for the thesaurus. This step involves selecting valid approximate duplicates for reference terms from the list of all approximate duplicates suggested by the computation. While this is a manual step, the amount of work required diminishes as time goes on since use can be made of prior manual efforts. Only new approximate duplicates discovered for reference terms need be presented to the domain expert for examination.

6.5.2   Sentence Identifier

The second stage of the methodology, *application-driven* cleaning, not only depends on the particular application, *summarization* in this case, but also on the domain at hand (e.g., case documents). For selecting the relevant sentences that will constitute the excerpt, first the sentences that make up the document have to be identified, which is the task of the *sentence identifier*. This process is relatively straightforward for clean documents that have normal punctuation. It is simply a matter of following standard punctuation rules. The process is made more difficult in dirty text since punctuation is haphazard; not only is there often a lack of punctuation marks as sentence delimiters, but also punctuation marks are often used for purposes other than punctuation. Furthermore, the presence of lines of programming code and tabular data make matters worse. Thus we remove them prior to identifying sentences by using a set of heuristics that we describe next.

Code Line Remover

Removal of lines of programming code from documents is an important step in identifying sentences from a document. In our case of customer support documents, lines of programming code may appear anywhere in the document, and they may appear frequently.

While not foolproof, we have found a simple approach to code removal to be quite effective. The approach is based on what might be called the *signature* of lines of programming code. That is, it is based on the characteristics that distinguish lines of programming code from normal lines of text. These characteristics include shorter lines, the presence of special characters unique to code (e.g., +=), and the presence of certain keywords (e.g., elsif). It is also necessary for there to be a minimum number of consecutive lines identified as code lines before any line is considered to be a line of programming code. This is due to the assumption that lines of programming code appear in blocks of lines rather than singly, which often occurs in practice. The special characters and keywords used to identify lines of programming code are kept in files that may be edited to reflect different programming languages or domains for the code lines. The minimum number of consecutive code lines required is a parameter to the code removal module.

Table Remover

Just as it is important to remove code lines when identifying sentences from a document, so is it important to remove tabular information. As with code line removal, the removal of table lines is based on the signature of a table. This signature includes such things as lines that have multiple spaces between words, lines whose words align in columns with those in succeeding lines, and multiple consecutive lines with these characteristics. The alignment of words or groups of words between lines may occur at the beginning of the word, the middle of the word, or the end of the word or group of words. So the document is parsed and information about the location, position, and length of words in each line is obtained.

The removal of table lines is simply a matter of identifying a minimum number of consecutive lines having these table characteristics. This minimum number of lines, as well as the number of spaces required between groups of words on a line and the minimum number of columns in the table are all parameters to the table removal module.

Sentence Boundary Identifier

Once code and tables have been removed, sentence boundaries are identified. This is a step that precedes the extraction of relevant sentences and therefore belongs to the *content extraction* stage of the methodology. Our approach to sentence boundary identification is a simple nonlinguistic approach that applies the standard punctuation rules when it can and resorts to other means of identification when it cannot. Thus, in addition to looking for sentence-ending punctuation (period, question mark, exclamation point) followed by a word starting with an upper case letter, we look for relaxed forms of these patterns as well as such things as header lines and blank lines to indicate end-of-sentence. We also set a limit on the number of lines that a sentence may occupy. If that limit is exceeded without finding an end to the sentence, the sentence is ended at that point and a new one is started. In our experience this set of simple heuristics helps prepare the documents to the point where sentence scoring can be more appropriately applied.

6.5.3   Sentence Extractor

Once the sentences of a document have been identified, their relevancy is analyzed. There are different aspects of a sentence that are indicative of its relevance. The words in a sentence, its correlation to other sentences, and its location are complementary aspects that we use for assessing the relevancy of a sentence. Thus the problem is reduced to identifying salient words or keywords and salient word correlations. Classical surface-level techniques that consider word occurrences, word location, and cue words are used to generate a keyword glossary that will be used to score a sentence according to its density of keywords. The technique used to score sentences according to their word correlations is an entity-based technique where the classical IR technique for computing the similarity among documents is scaled down to a single document. The individual or *local* scores obtained by each technique are combined into a weighted function that computes the *global* score of the sentence:

$$\text{global\_score}(\text{sentence}_i) = \sum_j (k_j \times \text{local\_score}_j(\text{sentence}_i)) ,$$

where k_j is the weight assigned to technique$_j$ according to its importance in the final scoring, local_score$_j$ represents the local score obtained by technique$_j$ and j can take any value in the set {"keyword", "semantic_similarity", "location"}.

The global score indicates the degree of relevance of a sentence and is used to select the m most relevant sentences that will be extracted into an excerpt. m is the number of sentences corresponding to the level of compression desired.

Next we describe some techniques that we use to compute the local scores.

Keyword Generator

A *glossary of keywords* is used to compute sentence scores as a weighted count of the keywords in the sentence divided by the sentence length. An entry in the glossary has the form (keyword, {indicator}), where {indicator} is a set of indicators of the techniques that identified that keyword. (Note that a keyword may be identified by more than one technique.) Since not all the keyword techniques are equally important, different weights that affect the sentence score computations are assigned to them. If a word has been identified by more than one technique, its weight as a combination of the weights of the corresponding techniques is increased. The local score, therefore, is computed as

$$\text{local\_score}_{\text{keyword}}(\text{sentence}_i) = \sum_j w_j(\text{keyword}_{ij}) ,$$

where w_j is a function of the weights of the techniques in the {indicator} set of keyword$_i$ entry, and keyword$_{ij}$ is keyword$_j$ in sentence$_i$.

The setting of these weights is done by experimentation, which makes it possible to adjust the weights to the characteristics of the output desired. If a large set of samples exists, then statistical methods could be used for this purpose. In the experiments done so far, we gave values of 1.0 to each weight,[7] but further experimentation will help us assess the relative importance of each technique for our current domain. The techniques used for generating the Keyword Glossary are now described.

Thematic Keywords

This technique is based on the intuitive notion that terms that appear frequently in the text and are not common across the collection are distinctive terms with the highest resolving power. Thus the density of keywords in a sentence gives a measure of the importance of the sentence. For computing the term scores, we use the standard TF-IDF[8] method, whereas Luhn [Luh58] only used TF:

$$\text{score}(\text{term}_j) = \text{TF}_j \times \log(100 \times n \times \text{IDF}_j) \, .$$

To choose the terms that are keywords a term score threshold has to be set. Again, the optimum value for this threshold is established by experimentation or by statistical methods. The keywords whose scores meet or surpass the threshold go into the glossary of keywords.

Location Keywords

This is a surface-level technique that follows the assumption that important sentences contain words that are used in important locations. Edmundson considered the title and headings of a document as the important locations and assumed that words in those locations were positively relevant [Edm68]. However, this assumption does not hold for all kinds of documents, in particular, those with generic section headings such as *cases,* where headings only indicate the kind of action logged in the corresponding section (e.g., "phone note"). To deal with this kind of document, we have extended Edmundson's technique to consider as important locations the actual content of sections with important headings, which is not the same as considering the headings themselves as important locations. For this purpose, we introduce the concept of *section model* which captures the names of important section headings. In doing so we are exploiting the fact that in some genres, regularities of discourse structure mean that certain sections tend to be important.[9] Such sections are assumed to have positively relevant words (i.e., non-stop-words *located* in those sections are keywords) that are entered in the glossary. However, when the list of non-stop-words coming from those sections is long, we can discard those that rarely appear in the document.

Cue Phrases

The Cue method was proposed by Edmunson [Edm68] under the hypothesis that the probable relevance of a sentence is affected by the presence of pragmatic

---

[7]A value of 0 for a weight disables the corresponding keyword technique.

[8]The standard connotation of term frequency is the frequency of the term within a document.

[9]Note that this is independent of their positions in the text.

words such as "significant" and "impossible." These words or phrases can be classified as *bonus* words, which are positively relevant and *stigma* words, which are negatively relevant. The method is domain dependent and requires manual inspection to identify cue phrases when they cannot be automatically generated with methods such as the one reported in [HL99]. The Keyword Glossary stores these phrases with the value "bonus" or "stigma" in the indicator field since bonus words have a positive weight while stigma phrases have a negative one.

Word Correlation: Semantic Similarity

This is an entity-level technique that discovers similarity relationships among sentences. It is inspired by the work on automatic text structuring and summarization reported in [SSMB99]. The idea is to apply classical IR techniques used for inter-document link generation in most hypertext link generation algorithms to generate intradocument links between sentences of a document. The intradocument links represent semantic similarity relationships given by the vocabulary overlap between the sentences. These relationships are obtained by computation of similarity coefficients between each pair of vectors of weighted terms representing the sentences. If the similarity between two vectors is large enough to be regarded as nonrandom, the vocabulary matches between the corresponding sentences are considered meaningful and the two sentences are semantically related. In this case, a semantic link between them is added. The number of links of a sentence is used to measure its centrality. The assumption is that well-linked sentences convey the contents of a document well. Therefore, the score assigned to a sentence by this technique is proportional to the number of its links:

$$\text{local\_score}_{\text{semantic\_similarity}}(\text{sentence}_j) = \text{COUNT}(\text{link}_{ij}) \,,$$

where link_{ij} represents a link from sentence_i to sentence_j.

In the current implementation we use the cosine coefficient as a metric of similarity. We take advantage of the existence of the Keyword Glossary (Section 6.5.3 – Keyword Generator) and use its keywords as the dimensions for the vector representation of sentences.

A few more techniques were implemented and experimented with, but they are not described here. It is important to recall the fact that the choice of techniques was limited due to the poor grammar, nonnarrative nature, and lack of rigorous discourse structure of case documents. The parameters of the techniques used can be manually set after experimentation or learned from a corpus of examples if such a corpus exists (which was not our case) as originally proposed in [KPC95].

6.6   Experimental Results

The FlexSum toolbox was applied to 5000 case documents to automatically generate excerpts. Analyzing the results implied a huge amount of work because for each document we had to have created an excerpt manually. So, we did a limited evaluation on 15 cases randomly picked. We set two levels of evaluation: to assess the performance of FlexSum as a tool to extract *technical* content from cases; and to assess its performance to extract *relevant* content. Figure 6.9 shows the results obtained for the case shown in Figure 6.8 using a default configuration of parameters for the different thresholds and weights.

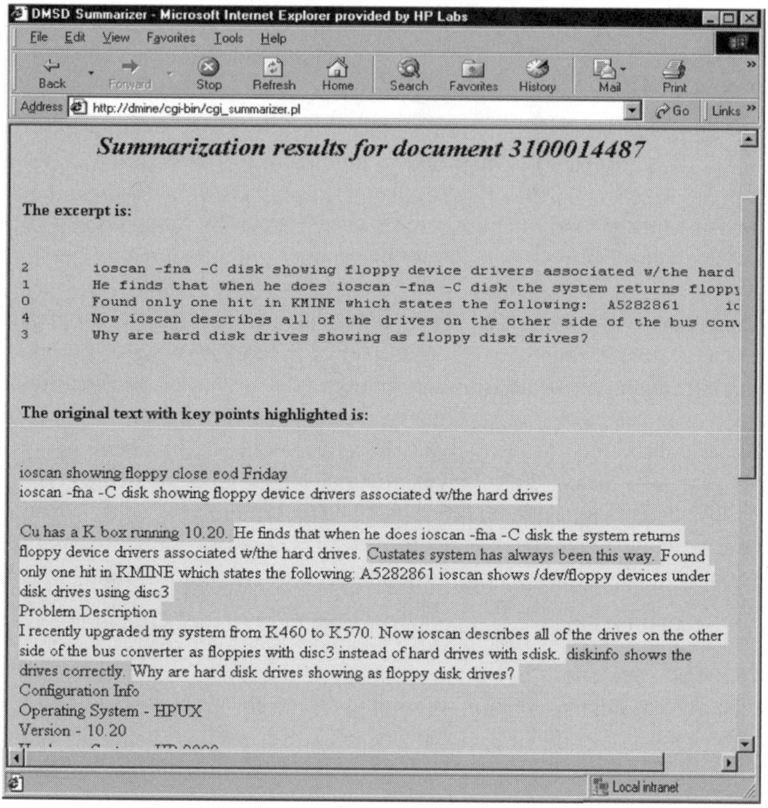

Figure 6.9. FlexSum results for case shown in Figure 6.8.

This figure corresponds to the browser page created by the visualization module of the summarizer with the results obtained after the sentence-scoring module ranked the sentences. In this page we can see the excerpt that is generated with the five most relevant sentences along with their ranking position preceding each of them. Following the excerpt is a portion of the highlighted version where the five most relevant sentences are highlighted on the original document. We can see that the five sentences found as the most relevant ones are of a technical nature. Although the portion shown in Figure 6.9 does not contains logistic sentences such as "called and left message but it was not Ted's number," the document does contain some, but all of them had such low relevance scores that the probability of being part of an excerpt is almost discarded. Furthermore, the five highest-ranked sentences obtained by combining the scores of the sentence extraction techniques are the same as those obtained in the excerpt generated manually. Of course, this was not always the case and we needed to evaluate the results on a test set which we obtained by randomly picking some cases and filtering out those that were not intuitive to us.

Since we wanted to evaluate whether any individual technique or the combination of all of them performed the best for the case domain, we computed for each alternative the average F-measure on the test set, finding that the combination technique was the best performer followed by the Semantic Similarity technique. Future work will evaluate the results on more cases and experiment with different settings of the parameters to see if we can obtain better excerpts for those case documents for which the current excerpts are not so good. Since one important feature of FlexSum is its flexibility in handling different application domains, we want to experiment with other domains as well. Also, automatic means to set the optimal values of the parameters for a particular application are to be researched as to the moment they are set manually by trial and error.

6.7   Mining Case Excerpts for Hot Topics

Once excerpts with the five most representative sentences of the cases are obtained, they are used to mine hot topics of reported problems. The assumption is that these sentences convey the essence of the problems reported in case documents. We cannot use the original cases because in addition to often being too long, they usually contain a lot of noise. Focusing only on the titles of the cases is not possible because often they are not representative of the problem and sometimes they are even left blank. Focusing on particular sections of the cases does not work either because cases are unstructured and sentences that describe the problem may be found anywhere. This is why in our approach we first mine for the most representative sentences and then use them to mine for hot topics.

The same steps used to mine the search log are used to mine the excerpts, except for the postfiltering step, because here we have only one view of the cases (the content view) so it is not possible to improve the quality of the hot topics by filtering out what another view would identify as extraneous documents. The first step is to transform the excerpts to vectors as described in Section 6.3.3 – Data Transformation, except that here for selecting the vector features we use the highest frequency words on the keyword glossary (we take advantage of the work already done by the keyword techniques of the sentence scorer in selecting relevant features) that are common to at least two excerpts. The next step is to apply a clustering algorithm on the vectors (we use SOM for facilitating visualization of the relationships between topics, but any other clustering algorithm could be used) as explained in Section 6.3.2. Finally, we do the labeling according to Section 6.3.4 to select as label the words that are common to at least half of the vectors in a cluster. We discard the topics for which a label cannot be found and from the ones remaining we consider to be hot those that have a minimum of m cases.

For validating the results, we face the same problem as for validating the hot topics of self-solving customers: the need for domain knowledge. We need to understand the content of the cases to determine whether they semantically belong to the topic to which their excerpts were assigned. As a result it is very labor intensive. Thus we did only a limited validation by randomly picking a few hot topics and computing the precision. (We cannot use the F-measure because we cannot compute the recall, given that we do not know which cases from the collection belong to a hot topic.) The average precision obtained was 75% and indeed much higher than using the original cases. A more extensive evaluation would be desirable.

6.8   Conclusions

Companies that are aware of the most common interests of their customers have a great advantage over their competitors. They can respond better and faster to their customer's needs, increasing customer satisfaction and gaining customer loyalty. In particular, for customer support it is essential to know what are the current hottest problems or topics that customers are experiencing in order to assign an adequate number of experts to respond to them, or to organize the website with a dynamic hot topics service that helps customers to self-solve their problems efficiently. Here we have presented an approach, implemented as the HotMiner toolbox, to automatically mine hot topics from the search and case logs of customer support centers. We have also shown some limited experimental results from applying it to Hewlett-Packard customer support logs. In contrast with other text mining work, our work deals with dirty/noisy text.

The novelty of our technique for mining hot topics from the search log is in its use to derive a search view of opened documents that is key to obtaining hot topics from the users' perspective. The other novelty is in using another view of the documents to boost the quality of the hot topics obtained from the first view by identifying extraneous documents resulting mainly from noisy clickstreams.

In our approach to mining hot topics from case logs we first obtain cleaner excerpts from the cases given that they are often very long and include a lot of noisy and irrelevant text. Sentences related to the problem are spread throughout a case so we need to extract the most relevant ones that capture the essence of the problem. Our approach to excerpt generation covers aspects of cleaning that are application independent, such as the resolution of typographical errors, as well as application-dependent aspects such as code removal that is considered a necessary step for sentence identification. Sentences are ranked according to the relevance scores given by different criteria embodied in a number of techniques that are suitable for nonnarrative grammatically poor text. Once the excerpts from the cases have been obtained, they are mined for hot topics analogously as for mining document search views.

We have performed a limited validation of the approach. The results are encouraging and show that the approach is capable of discovering hot topics from dirty text with an acceptable quality. Further experimentation and validation of the approach is still to be done and finding ways to automatically set an optimal configuration of parameters for a particular domain is planned for future work. Our wish list also includes the extension of this work to the proactive detection of epidemic behavior in customer problems. This work, among other things, gave us the opportunity to gain a good understanding of the characteristics of dirty text, the limitations that it imposes, and the opportunity to explore techniques to deal with it.

Acknowledgments: I want to thank Jim Stinger for his work on the implementation of the cleaning modules of FlexSum, as well as for always being there when I had an implementation problem. Part of the material on cleaning in the second part of the chapter was taken from an earlier paper that we coauthored. His editorial assistance improved the readability of the material presented here. I am also grateful to Umesh Dayal and Meichun Hsu for the valuable discussions, continuous feedback, and encouragement during the development of the work reported here; to Maria Eugenia Castellanos for her valuable suggestions on the use of statistics for the postfiltering stage; to Nina Mishra for her intellectual and motivational support; to Jaap Suermondt for carefully reviewing this chapter; and finally, to Manuel Garcia-Solaco, for his comments and assistance in converting the MS Word version to LATEX.

References

[BE99]   R. Barzilayand and M. Elhadad.*Using Lexical Chains for Text Summarization.*In [MM99], 1999.

[CHJ61]   W.D. Climenson, H.H. Hardwick, and S.N. Jacobson.Automatic syntax analysis in machine indexing and abstracting.*American Documentation*, 12(3):178–183, 1961.

[CKPT92] D.R. Cutting, D.R. Karger, J.O. Pedersen, and J.W. Turkey.Scatter/gather: A cluster-based approach to browsing large document collections.In *Proceedings of the Fifteenth Annual International ACM SIGIR Conference on Research and Development in Information Retrieval*, Copenhagen, Denmark, pages 318–329, Jun 1992.

[DHS01]   R.O. Duda, P.E. Hart, and D.G. Stork.*Pattern Classification*, second edition.Wiley, New York, 2001.

[DPHS98] S. Dumais, J. Platt, D. Heckerman, and M. Sahami.Inductive learning algorithms and representations for text categorization.In *Proceedings of the ACM CIKM International Conference on Information and Knowledge Management*, Bethesda, MD, Nov 1998.

[Edm68]   H.P. Edmundson.New methods in automatic extraction.*Journal of the ACM*, 16(2):264–285, 1968.

[EE98]   M.A. Elmi and M. Evens.Spelling correction using context.In *Proceedings of the 36th Annual Meeting of the ACL and the 17th International Conference on Computational Linguistics*, pages 360–364, 1998.

[HL99]   E.H. Hovy and H. Liu.The value of indicator phrases for automated text summarization.Unpublished, 1999.

[Inx]   Inx [online].Available from World Wide Web: www.inxight.com/ products/linguistx.

[KHKL96]   T. Kohonen, J. Hynninen, J. Kangas, and J. Laaksonen.Som_pak: The self-organizing map program package.*Laboratory of Computer and Information Science*, Report A31, 1996.

[Koh92]   T. Kohonen.*The Self-Organizing Map. Neural Networks: Theoretical Foundations and Analysis*.IEEE Press, New York 1992.

[KPC95]   J. Kupied, J. Piedersen, and F. Chen.A trainable document summarizer.In *Proceedings of the Eighteenth Annual International SIGIR Conference on Research and Development in Information Retrieval*, pages 68–73, 1995.

[Kue87]   G.H. Kuenning.International ispell version 3.1.00.ftp.cs.ucla.edu, 1987.

[Kuk92]   K. Kukich.Techniques for automatically correcting words in text. *ACM Computing Surveys*, 24(4):377–439, 1992.

[Leh82]   W.G. Lehnert.*Plot Units: A Narrative Summarization Strategy*.Erlbaum, Hillsdale, NJ, 1982.

[LSCP96]   D. Lewis, R. Schapire, J. Cllan, and R. Papka.Training algorithms for linear text classifiers.In *Proceedings of SIGIR-96, Nineteenth ACM International Conference on Research and Development in Information Retrieval*, 1996.

[Luh58]   H.P. Luhn.The automatic creation of literature abstracts.*IBM Journal of Research and Development*, 2(2), 1958.

[Mar97]   D. Marcu.*The Rhetorical Parsing, Summarization and Generation of Natural Language Texts*.PhD dissertation. University of Toronto, 1997.

[McI82]   M.D. McIlroy.Development of a spelling list.*IEEE Transactions on Communication, COM-30*, 1:91–99, Jan 1982.

[MM99]   I. Mani and M. Maybury.*Introduction. Advances in Automatic Text Summarization*.MIT Press, Cambridge, MA, 1999.

[Nun90]   G. Nunberg.The linguistics of punctuation.*Center for the Study of Language and Information Lecture Notes* 90(18), 1990.

[RR94]   J. Reynar and A. Ratnaparkhi.A maximum entropy approach to identifying sentence boundaries.In *Proceedings of the Conference on Applied Natural Language*, 1994.

[Sal89]   G. Salton.*Automatic Text Processing: The Transformation, Analysis, and Retrieval of Information by Computer*.Addison-Wesley, Reading, MA, 1989.

[SKK00]   M. Steinbach, G. Karypis, and V. Kumar.A comparison of document clustering algorithms.In *Proceedings of the KDD Workshop on Text Mining*, 2000.

[SSMB99]   G. Salton, A. Singhal, M. Mitra, and C. Buckley.*Automatic Text Structuring and Summarization*.In [MM99], 1999.

[Sti00]   J.R. Stinger.Automatic table detection method and system.HP Internal Paper, 2000.

[SW81]   T.F. Smith and M.S. Waterman.Identification of common molecular subsequences.*Journal of Molecular Biology*, 147:195–197, 1981.

[Too00]   J. Toole.Categorizing unknown words: Using decision trees to identify names and misspellings.In *Proceedings of the Sixth Applied Natural Language Processing Conference*, pages 173–179, 2000.

[Wil88]   P. Willet.Recent trends in hierarchical document clustering: A critical review.*Information Processing and Management*, 577(97), 1988.

[YY99]   X. Lui Y. Yang.A reexamination of text categorization methods.In *Proceedings of the 22nd International Conference on Research and Development in Information Retrieval (SIGIR'99)*, University of California, Berkeley, pages 42–49, 1999.

7

Combining Families of Information Retrieval Algorithms Using Metalearning

Michael Cornelson
Ed Greengrass
Robert L. Grossman
Ron Karidi
Daniel Shnidman

Overview

This chapter describes some experiments that use metalearning to combine families of information retrieval (IR) algorithms obtained by varying the normalizations and similarity functions. By metalearning, we mean the following simple idea: a family of IR algorithms is applied to a corpus of documents in which relevance is known to produce a learning set. A machine learning algorithm is then applied to this data set to produce a classifier that combines the different IR algorithms. In experiments with TREC-3 data, we could significantly improve precision at the same level of recall with this technique. Most prior work in this area has focused on combining different IR algorithms with various averaging schemes or has used a fixed combining function. The combining function in metalearning is a statistical model itself which in general depends on the document, the query, and the various scores produced by the different component IR algorithms.

7.1   Introduction

This chapter describes some experiments that use metalearning to combine families of information retrieval algorithms obtained by varying the normalizations and similarity functions. In experiments with TREC-3 data, we could significantly improve precision at the same level of recall with this technique.

In more detail, our goal is to satisfy a query q by returning a ranked list of documents d from a collection C, that are relevant to the query. Our point of view is that there is no single best algorithm for solving this problem. On the contrary, we assume that we have m algorithms $A_1, \ldots, A_m$ and that some work well for certain queries and certain sets of documents, while others work well for other queries and other document sets.

Our approach is to apply metalearning [Die97, PCS00] to learn how to select which algorithm A_j or combination of algorithms to use, where the selection is a function of the query q and the document d. In more detail, we assume that we are given a collection C of documents d and queries q for which we know the relevance; that is, for each document-query pair (d, q), C also contains a human judgment of whether d is relevant to q. This is the "training set" from which we will learn how to select the best algorithm(s). If we apply all the algorithms A_1, $\ldots, A_m$ to each document-query pair (d, q), this produces a learning set ML. ML denotes metalearning set, a terminology which we explain below. Our approach is to use machine learning to create a predictive model from ML that can be used on previously unseen queries and documents to determine which algorithm or combination of algorithms to use to determine relevance. We assume that each algorithm A produces a score $A(d, q)$ between 0 and 1 when applied to a document d and query q. The higher the score is, the more relevant the document. In general, the resulting ML will contain, for each document-query pair (d, q), a set of features characterizing d and q, the relevance score computed by each of the algorithms A_i, and the human relevance judgment or "truth."

We introduce this approach in this chapter and describe an experimental study that provides evidence regarding its effectiveness. We confine ourselves to the simplest case in which we use the scores $A(d, q)$ themselves and not features of the query or the document to determine which algorithm to use. (But it should be stressed that we do NOT use the relevance judgments in selecting an algorithm for a given test document. The model we build is genuinely predictive, not retrospective.) We will treat the more general case where we use the document and query features as the independent (predictor) variables in future work.

Our point of view is that metalearning provides a natural framework to combine similar or quite different IR algorithms. We believe that the following aspects of this work are novel.

1. Our metalearning approach allows us to select a distinct algorithm or combination of algorithms for each new document/query pair submitted to our metamodel. Most previous work in algorithm combination assumes a fixed combining function, which in most cases is simply an average of the scores.

2. Our approach allows us to create a combining function that summarizes variability across both documents and queries, while prior work known to us has aggregated data across documents alone.

3. We have investigated several nonlinear combining functions, while prior work known to us has investigated linear combining functions.

This chapter is organized as follows: Section 2 describes related work. Section 3 describes background work in information retrieval, and Section 4 describes background work in data mining. The background material in Sections 3 and 4 is provided so that researchers in the data mining community can easily read this chapter. Section 5 describes our implementation, and Section 6 contains our experimental results. Section 7 contains the summary and conclusion.

7.2   Related Work

There has been some prior work on combining information retrieval algorithms. A number of researchers have conducted experiments to study the effect and possible benefit of satisfying a given information need from a given document collection by combining the results of multiple IR algorithms. Typically, two [Lee95] to six [Lee97] algorithms have been combined. In most cases, each algorithm is applied to a given collection, for a given query, resulting in a ranked list of document scores for each algorithm. Then some function is evaluated that combines the scores computed by the different algorithms for a given document. Often, the scores are normalized before combination. The end result is a single ranked list of document scores. The combination formula is usually a function of the scores computed by the participating algorithms and (sometimes) the number of scores, that is, the number of algorithms that retrieved a given document. In most studies averages of one type or another are used to combine the scores.

Multiple algorithms can be obtained by using the same basic IR method, for example, the vector space method with cosine similarity, but executing the method with multiple term weighting schemes [Lee95, Lee97]. Instead of varying the term weighting scheme, one can vary the kind of terms used as document/query descriptors, such as words versus phrases. [MMP00]. Alternatively, different basic IR methods can be employed, for example, P-norm extended Boolean and vector similarity [FS94].

Hull et al. [HPS96] investigated combining three learning algorithms (nearest neighbor, linear discriminant, and a neural network) with one IR algorithm (Rocchio query expansion) by averaging the scores produced by the different algorithms.

Some limited theoretical analyses of these combination results have been performed. Lee [Lee95] studied the properties of several classes of the term weighting scheme, and offered reasons for expecting that each scheme would favor a different class of documents so that certain pairs of weighting functions, each drawn from a different class should perform better than one weighting function alone. Vogt and Cottrell [VC98] studied the properties that IR systems should possess in order that linear combinations (i.e., linear combinations of their normalized scores) should produce a better document ranking than either system by itself. Their results were limited to systems in linear combinations. The performance measures they use as predictors of effective combined performance include both measures of individual systems (e.g., average precision) and pairwise measures (e.g., Guttman's Point Alienation (GPA), a measure of how similar the document rankings produced by two systems are to each other).

On rare occasions, an effort is made to develop a metamodel, that is, to select the best IR model(s) for a given document. For example, Mayfield et al. [MMP00] ran a test where the document and query vectors were based on either words or phrases, depending on the document to be evaluated. They found that for some queries, this strategy produced a significant gain, while for others it did not or actually degraded performance. Unfortunately, they did not develop a metamodel that would accurately predict when the strategy should be employed [May00].

7.3   Information Retrieval

For completeness and to fix notation, in this section we briefly review some standard definitions and techniques in information retrieval (IR) for those not familiar with this subject. For additional information, see [Gre01].

Setup

Given a corpus of C documents d and a query q, the goal is to return the documents that best satisfy the query. Assume that r of the p documents returned by the IR system are relevant to the query and that overall R of the documents in the corpus C are relevant to the query. Typically, the IR system returns a list of documents, ordered (ranked) according to the probability or degree of relevance to q, as measured by some system metric. The system may return, or the user may examine, all documents for which the system-computed metric exceeds a specified threshold. Let p be the number of documents above this threshold. The *recall* is defined to be r/R, while the *precision* is defined to be r/p. The goal is to find algorithms with high precision or high recall, or some desired trade-off between the two.

Vector Space Method

Rather than work with the documents directly, we define a feature vector $F(d) \in R^n$ for each document d, as is standard [Sal89]. If we view a query as a document, then in exactly the same way we can define a feature vector $F(q)$ of the query q. Fix a relevance measure

$$m(u, v) : R^n \times R^n \longrightarrow [0, 1]$$

that ranges from 0 to 1, with 1 indicating that u and v are highly relevant and 0 indicating that they are not. Given a query q, our strategy is to return the p documents d such that $u = F(d)$ is closest to $v = F(q)$ with respect to the measure $m(\cdot, \cdot)$. These are simply the p-highest scoring document-query pairs.

Document-Term Matrix

A document-term matrix is created from a collection of documents and contains information needed in order to create a feature vector $F(d)$ for a document d. From the collection C of documents, we create what is called a dictionary by extracting terms, where the terms are used to characterize the documents. For example, the terms can consist of words in the collection, phrases in the collection, or n-grams in the collection. An n-gram consists of n-consecutive characters, including white space and punctuation. From these data, we form a document-term matrix $d[i, j]$, where, at the simplest, $d[i, j]$ represents the number of times the term j occurs in document i. Usually, a weighting is used to normalize the terms in the document-term matrix.

In practice, not all terms end up in the dictionary. Certain common terms (e.g., articles, prepositions, etc.) occur so widely in the corpus that they have little value for classifying document relevance relative to a given query. Hence such words are placed in a "stop-list." Words in the stop list are excluded from the dictionary. Also, it is common to use a stemming algorithm to reduce words into a common root and only include the root in the dictionary. For n-grams, stop-lists and stemming are inapplicable, but statistical techniques can be employed to eliminate n-grams having little value for predicting relevance.

Normalization

In Table 7.1 below, we list some of the common term weighting functions (normalizations) used. In our metalearning experiments, we vary these normalizations. In the table tf = number of times the term appears in the document, wc = total number of terms in document, N = total number of distinct terms in dictionary, IDF is the Inverse Document Frequency (a measure of how rare a term is in C), TF is the frequency of the most frequently occurring term in the given document, and n = number of times that term occurs in collection C.

Similarity Scores

In Table 7.2 below, we list some common similarity metrics found in the literature, that is, metrics used to measure the similarity of two feature vectors, FV_1 and FV_2. Distance metrics d have been converted to similarity metrics by the transformation $s = (d + 1)^{-1}$ [vR79]. In our metalearning experiments, we vary these similarity scores. In the table n_1 equals the number of nonzero terms in FV_1, n_2 equals the number of nonzero terms in FV_2, and w equals the number of terms common to FV_1 and FV_2. Also z equals the number of distinct terms that are neither in FV_1 nor in FV_2, and N equals the total number of distinct terms in the FV space. Therefore, $N - z$ equals the total number of distinct terms that occur in FV_1 or FV_2 or both.

term frequency	tf
word count	tf/wc
log	$\log(tf + 1)$
binary	$0 \, or \, 1$
maximum normalization	tf/TF
average normalization	$tf/(TF - \mathrm{avg}(tf))$
TF*IDF	$tf * (N/n)$
TF*ln(IDF)	$tf * ln(N/n)$

Table 7.1. Some Common Normalizations

Euclidean	$1/[sqrt(\sum((FV1_i - FV2_i)^2))) + 1)]$
City Block	$1/[\sum(FV1_i - FV2_i)) + 1)]$
Maximum	$1/[\max(FV1_i - FV2_i)) + 1)]$
Dice's Coefficient	$2 * w/(n_1 + n_2)$
Jaccard's Coefficient	$w/(N - z)$

Table 7.2. Some Common Similarity Functions

Feature Vector	Truth
x_1	y_1
$\vdots$	$\vdots$
x_k	y_k

Table 7.3. A Learning Set

7.4 Metalearning

For completeness and to fix notation, in this section we briefly review some standard definitions from metalearning for those not familiar with this subject. For additional information, see [Die97].

Supervised Learning

In supervised learning, we are given a data set of pairs (x, y) called a learning set. Commonly, $x \in R^n$ is an n-dimensional feature vector containing the independent variables or attributes, and $y \in \{0, 1\}$ is the dependent variable or truth (see Table 7.3). The goal is to construct a function or model f,

$$y = f(x) = f_a(x) = f(x; a), \qquad a \in A,$$

where $f = f_a$ is defined by specifying parameters $a \in A$ from an explicitly parameterized family of models A [GL02]. There are well-known parameterizations for a wide variety of statistical models, including linear regression, logistic regression, polynomial regression, generalized linear models, regression trees, classification trees, and neural networks.

Fix a model $f = f_a$. The model is applied to previously unseen feature vectors x (i.e., feature vectors outside the learning set) to predict the truth $y = f(x)$. We measure success by using a test set (see Table 7.4) distinct from the learning set and measuring the deviation of $f(v)$ from t over the test set.

Feature Vector	Predicted Truth	Truth
v_1	$f(v_1)$	t_1
$\vdots$	$\vdots$	$\vdots$
v_j	$f(v_j)$	t_j

Table 7.4. A Test Set

Metalearning

Broadly speaking, learning is concerned with finding a model $f = f_a$ from a single learning set, while metalearning is concerned with finding a metamodel $f = f_a$ from several learning sets $\{L_1, \ldots, L_n\}$, each of which has an associated model $f = f_a[j]$. The n component models derived from the n learning sets may be of the same or different types. Similarly, the metamodel may be of a different type than some or all of the component models. Also, the metamodel may use data from a (meta-) learning set, which are distinct from the data in the individual learning sets L_j.

We begin by describing two simple examples of metalearning. For the first example [GBNP96], we are given a large data set L and partition it into n disjoint subsets $\{L_1, \ldots, L_n\}$. Assume that we build a separate classifier on each subset independently to produce n classifiers $\{f_1, \ldots, f_n\}$. For simplicity, assume that the classifiers are binary so that each classifier takes a feature vector and produces a classification in $\{0, 1\}$. We can then produce a metamodel simply by using a majority vote of the n classifiers.

As the second example, given a learning set L, we replicate it n times to produce n learning sets $\{L_1, \ldots, L_n\}$ and create a different model f_j on each learning set L_j, for example, by training the replicated data on n different model types. Given a feature vector x, we can produce n scores $(f_1(x), \ldots, f_n(x))$, one for each model. Given a new learning set ML, we can build a metaclassifier f on ML using the data $\{(f_1(x), \ldots, f_n(x), y) : (x, y) \text{ in } ML\}$.

In the work below, the f_j are different information retrieval algorithms trained on the same learning set L as in the second example. We combine them using a polynomial or other basic model. The combining model uses the scores produced by the base algorithms and not any document or query vector features directly.

7.5   Implementation

In this section, we briefly describe the system we used for these experiments. The experiments were done using a text mining module added to the PATTERN [PAT99] data mining system.

PATTERN is designed around ensemble-based learning. That is, the basic structure in the system is an ensemble of models. Ensembles in PATTERN naturally arise by partitioning the learning set, by sampling the learning set, and by combining multiple classifiers. For the experiments described below, we simply viewed the outputs of experiments involving different IR algorithms as a metalearning set. We then used PATTERN to compute a (meta-) classifier.

The PATTERN text mining module creates a dictionary and document-term matrix by scanning a collection of documents. The dictionary can be composed of either words or *n*-grams. In either case, stop-lists can be applied to remove common terms. In the case of words, a stemming algorithm can also be applied.

PATTERN's document-term matrix and feature vectors use sparse data structures, since several of the experiments use over 150,000 terms. This size of the feature vectors was a problem for PATTERN initially and some effort was required for PATTERN to work with feature vectors of this size.

7.6   Experimental Results

We created a learning set using FBIS documents from the TREC [Har95] collection and estimated the parameters for four different combining functions: linear regression, two different quadratic regressions, and cubic regression.

For the validation, we used 300 FBIS documents and 29 queries. This produced 8700 query document combinations. One hundred and ten of these combinations were classified as relevant in the TREC data set; that is, in 110 of the 8700 query-document combinations, the query was labeled as relevant to the corresponding document. For each document-query pair, we computed $40 = 8 \times 5$ scores, by using the five similarity metrics and the eight term weight normalizations from Tables 7.1 and 7.2. These 40 scores were the independent variables used for each combining function. To simplify the interpretation of our experiments, we selected one of the best performing IR algorithms as a baseline. The second column of Table 7.5 represents the number of documents required by the baseline algorithm in order to obtain the number of relevant documents given in the first column. The remaining columns give the number of the documents required by the different combining methods (linear regression, cubic regression, and two different quadratic regressions) in order to obtain the number of relevant documents indicated. The 40 scores generated for each query-document pair were combined using each of the four combining methods to obtain a single combined score for the given pair. The query-document pairs were ordered by these combined scores.

No. Relevant	Baseline	Linear	Cubic	Quad. 1	Quad. 2
10	592	380	145	75	76
20	1352	931	299	212	195
30	1825	1417	820	400	372
40	2453	1820	1122	670	622
50	4013	2306	1422	933	920
55	4383	2442	1598	1168	1137
60	5044	2601	1730	1458	1345
70	5817	3136	2101	1910	1734
80	6733	4164	2775	2668	2312
90	7420	4880	3472	3645	3035
100	8328	5552	5211	4875	5353
110	8644	7597	8101	7780	7948

Table 7.5. Results of Applying Four Different Metalearning Algorithms to the 40 Algorithms

As can be seen from Table 7.5, the various combining functions significantly outperformed the baseline. Note that the baseline did not perform particularly well on this data set. Only very simple term weighting and similarity metrics were employed in this proof of concept, including some that are found in the literature, but are rarely used today. This points out that combining even relatively weak IR algorithms using metalearning can produce an IR algorithm that is strong enough to be useful.

Note that by looking at the first row in Table 7.5 , we see that the recall for the baseline algorithm is $10/110 = 0.090$ and the precision is $10/592 = 0.016$. From the same row, we see that for the quadratic combining function and the same recall, the precision is $10/75 = 0.133$. That is, for the same recall, the nonlinear quadratic combining function increases the precision by over 700%.

7.7   Further Work

The work described in the body of this chapter has provided encouragement for pursuing and greatly expanding the metalearning approach to IR. Below, we briefly describe the directions and successes that this expanded effort has taken. A further paper, to be published in the near future, will describe this expanded effort in much greater detail.

The study was expanded from the preliminary 300 to 11,188 FBIS documents, and 10 queries, chosen to have reasonable but not excessive representation in the document set. The number of term weighting schemes and document/query similarity metrics was expanded, resulting in 54 possible IR algorithms; then algorithms that made no sense or were too closely correlated to earlier algorithms were eliminated, resulting in 28 IR algorithms, ranging from state of the art to found only in textbooks to homegrown.

The combining functions described in the body of this chapter were replaced by metalearning functions drawn from the Machine Learning (ML) world, for example, decision trees, logistic regression, and later Random Forests (RF), a novel tree-ensemble method developed at Berkeley. The results using these ML metamodels confirmed the results reported in this preliminary study. Retrieval using a metamodel trained on a variety of simple IR models significantly outperforms any of the individual IR models taken separately. We can express these results in slightly different terminology: The collection of IR algorithms combined by the metamodel is an ensemble. The individual IR models are base models. We found that an ensemble could consistently outperform a varied collection of simple base models. Interestingly, the base models that contributed most to the performance of the ensemble were not necessarily the models developed from the algorithms most recommended in the IR community.

7.8   Summary and Conclusion

To summarize, in this chapter, we have introduced the basic idea of using metalearning to combine several different information retrieval algorithms to improve precision and recall. The idea of metalearning is simple: a family of IR algorithms applied to a corpus of documents in which relevance is known produces a learning set. A machine learning algorithm applied to this set produces a classifier that combines the different IR algorithms.

We have also presented evidence of the effectiveness of this approach when simple nonlinear combining models are used on a collection of IR algorithms produced by varying the similarity and normalizations.

The experimental study described here is promising, but additional work is required to understand the applicability of this approach to larger and more heterogeneous data sets and using a wider variety of IR algorithms.

Acknowledgments: This work was supported in part by the CROSSCUT Program from DOD and in part by Magnify.

References

[Die97]   T.G. Dietterich.Machine-learning research: Four current directions.*AI Magazine*, 18(4):97–136, 1997.

[FS94]   E.A. Fox and J.A. Shaw.Combination of multiple sources.In *Proceedings of the Second Text Retrieval Conference (TREC-2)*, pages 97–136, 1994.

[GBNP96] R.L. Grossman, H. Bodek, D. Northcutt, and H.V. Poor.Data mining and tree-based optimization.In *Proceedings of the Second International Conference on Knowledge Discovery and Data Mining*, E. Simoudis, J. Han and U. Fayyad, eds., AAAI Press, Menlo Park, CA, pages 323–326, 1996.

[GL02]   R.L. Grossman and R.G. Larson.A state space realization theorem for data mining. In subm., 2002.

[Gre01]   E. Greengrass.Information retrieval: A survey.United States Department of Defense Technical Report TR-R52-008-001, 2001.

[Har95]   D.K. Harman, editor.*Proceedings of the Third Text Retrieval Conference (TREC-3)*. National Institute of Standards and Technology Special Publication 500-226, 1995.

[HPS96]   D.A. Hull, J.O. Pedersen, and H. Schütze.Method combination for document filtering.In *Proceedings of the Nineteenth Annual International ACM SIGIR Conference on Research and Development in Information Retrieval*, 1996.

[Lee95]   J.H. Lee.Combining multiple evidence from different properties of weighting schemes.In *Proceedings of the Eighteenth Annual International ACM SIGIR Conference on Research and Development in Information Retrieval*, 1995.

[Lee97]   J.H. Lee.Analyses of multiple evidence combination.In *Proceedings of the Twentieth Annual International ACM SIGIR Conference on Research and Development in Information Retrieval*, 1997.

[May00]   J. Mayfield.Personal communication, 2000.

[MMP00]   J. Mayfield, P. McNamee, and C. Piatko.The JHU/APL HAIRCUT System at TREC-8.National Institute of Standards and Technology Special Publication, 2000.

[PAT99]   PATTERN.The pattern system version 2.6, Magnify, Inc., 1999.

[PCS00]   A.L. Prodromidis, P.K. Chan, and S.J. Stolfo.Meta-learning in distributed data mining systems, issues and approaches.In *Advances in Distributed Data Mining*, Hillol Kargupta and Philip Chan, eds., MIT Press, Cambridge, MA, pages 81–113, 2000.

[Sal89]   G. Salton.*Automatic Text Processing: The Transformation, Analysis, and Retrieval of Information by Computer*.Addison-Wesley, Reading, MA, 1989.

[VC98]   C.C. Vogt and G.W. Cottrell.Predicting the performance of linearly combined IR systems.In *Proceedings of the 21st Annual International ACM SIGIR Conference on Research and Development in Information Retrieval*, pages 190–196, 1998.

[vR79]   C. J. van Rijsbergen.*Information Retrieval*, second edition.Butterworths, London, 1979.

Part III

Trend Detection

8

Trend and Behavior Detection from Web Queries

Peiling Wang
Jennifer Bownas
Michael W. Berry

Overview

In this chapter, we demonstrate the type and nature of query characteristics that can be mined from web server logs. Based on a study of over half a million queries (spanning four academic years) to a university's website, it is shown that the vocabulary (terms) generated from these queries do not have a well-defined Zipf distribution. However, some regularities in term frequency and ranking correlations suggest that piecewise polynomial data fits are reasonable for trend representations.

8.1   Introduction

The Web has made end-users' searching a reality with a mixed blessing. The vast amount of information on the Web is readily available to those who have Internet access and know how to find it, but the majority of Web users show perpetual novice searching behaviors. Much has been done to improve the functionality of the search engines in the last two decades. Adding sophisticated features to search engines can certainly enhance systems' utility, but only sophisticated users can benefit from the advanced functions. To improve usefulness and usability, however, research is still needed in the interpretation of Web users' behavior.

Studies of traditional information retrieval (IR) systems (such as library catalogs and various online databases) reveal many problems that searchers encountered: the complexity of query syntaxes, the semantics of Boolean logic operators, and the linguistic ambiguities of natural language. These studies made suggestions on possible improvements in system design as well as user training. The current Web users, however, are unlikely participants in user training programs given the fact that anyone can get a quick start on interacting with the *friendly* systems.

Therefore, the system must be designed to facilitate self-learning and intelligent interactions beyond the traditional IR systems.

Several studies have reported results and analysis of collected queries and logged search sessions from Web search engines [JSS00, SHMM99, SWJS01, WP97]. A recent publication reviews studies on Web searching and suggests future research [JP01]. *Excite* search logs were collected twice: 51,473 queries on March 9, 1997 and 1,025,910 queries on September 16, 1997 [JSS00, SHMM99]. *AltaVista* search logs recorded 993,208,159 queries from August 2 to September 13, 1998 [WP97]. A study by Hoelscher captured 16,252,902 queries submitted to the *Fireball* search engine in German from July 1 to July 31, 1998 (cited in [JP01]). Despite the differences in data collection and processing as well as the foci of analyses, results are comparable in many aspects: Web queries are short, averaging two words, and very simple in structure; few queries use advanced search features; and many queries contain syntax or semantic errors, resulting in zero-hits (33%) [SBC97, WP97]. Term association has been examined by *Excite* and *AltaVista* researchers using quite different parsing strategies. Both groups report that term pairs do not follow a Zipf frequency-ranking distribution; some of the high-frequency pairs are not topic-related, such as *and-and, of-the, how-to, and-not*, and so on; strong association exists among certain topic terms such as *cindy-crawford, visual-basic* [RW00, SWJS01, Wol99].

Although the logs in these studies were quite large, they only covered a very short period of time and hence provide only a brief snapshot of Web searching. This study reports on an analysis of 541,920 logged queries from May 1997 to May 2001 at a university's main website. These particular data facilitate the analysis of longitudinal trends and changes over time.

8.2   Query Data and Analysis

A total of 541,920 queries were submitted to a search engine at the University of Tennessee's website from May 1997 to May 2001. Although a new search engine was added to the website in January 2000, the original and new search engines co-existed for more than a year until the final retirement of the old engine.[1] No log file was available from the new (interim) search engine. With regard to completeness, our data set includes complete queries covering two academic years (August 1997 to July 1999) or two calendar years (January 1998 to December 1999).

The log data include a date stamp, the query statement, and the hits (number of URLs retrieved per query). Each query is assigned a unique identification number for storage in a relational database. The following are selected examples showing some interesting characteristics of the Web queries.

---

[1] Built using SWISH, a shareware program.

Unique Id	Date Stamp	Query Statement	Hits
3	1997/05/06	summer '97	0
18490	1998/09/21	"423-525-8438"	0
32595	1997/09/16	ALUMNI HEADQUARTERS FOR 10/18/97 TENNESSEE VS˙ ALABAMA FOOTBALL GAME AT BIRMINGHAM	0
55389	1997/11/11	the beacon	0
55390	1997/11/11	the daily beacon	0
63151	1997/12/02	Drop out rate for Freshmen at the University of Tennessee in the 1996-1997 school year	0
249599	1999/01/05	Anthropology	50
507033	2000/09/05	"how much garbage does the university of Tennessee at Knoxville dispose of in one month?"	0
507046	2000/09/05	3270	0
535087	2001/03/13	ewan AND unix AND domain AND help AND setup	0
541736	2001/05/10	tennesee 101 fall	0

The raw data were parsed, cleaned, and then imported into ACCESS, a relational database. Data mining in this context is focused on both statistical and linguistic trends in these data. The statistical analyses provide a global view of search activities over time. Linguistic analyses examine queries as statements, vocabulary, and word associations.

Word association is viewed as the co-occurrence of two words, called a *word pair*. A word pair from a query is the adjacent words ignoring word order, or two words in the proximity of one intervening word. This decision was based on the fact that about 54% of the queries contained two or three words and 6% of the queries had more than three words.

8.2.1   Descriptive Statistics of Web Queries

Of the 541,920 queries collected, 73,834 were from 1997, 172,492 were from 1998, 233,442 were from 1999, 43,448 were from 2000, and 18,704 were issued in 2001. A substantial increase is seen from 1998 to 1999. The drastic decrease in number of queries was due to the search engine change in 2000 and the final retirement of the old engine in 2001 (mentioned earlier). Several observations were made about the query characteristics: zero hit, empty, length measured by the position count between the first and the last characters, the number of words in cleaned queries, and unique or common queries (identical in words and wording).

Zero-hit queries ranged from 32 to 40% with an average 33% for all queries. This number is much higher than those reported by the other studies of Web queries. Several factors contributed to the zero hit. First, the stop-words set by the engine excluded many words such as *the*, *of*, numeric, nonalphabetic symbols, and words with high frequencies. Although this fact was readily available on the search engine web page, users seldom checked it. Second, the queries contained a high percentage of misspelled words (26%). Third, many searches entered names of individuals. The name search is not handled by this engine since a *People Search* utility was provided for that particular need. In addition, only names appearing on the web pages that were three hyperlinks from the main source page were indexed. Fourth, the default Boolean operator was AND, thus a query containing three words was treated as a search statement using Boolean AND. To search using Boolean OR, the OR had to be included in the queries. Finally, empty queries also contributed to the zero-hit in a small scale (1.4%). Two previous studies reported a comparatively higher percentage of empty queries: 5% (*Excite* queries) and 20% (*AltaVista* queries).

The average length of the queries excluding empty queries was 13 positions between the first character and the last character. The longest one had 131 positions in natural language format with punctuation:

> "I have had this problem ever since the beginning of this week. I have not been able to dial into UTK network even with one success."

Most nonempty queries were short: about 40% were single word queries, 40% were two-word queries, 14% were three-word queries, and 6% of the queries contained four or more words. The maximum number of words in a query was 26 and the mean was 2, much shorter than those reported by other studies. For the set of cleaned queries, there were 135,036 unique queries, of which 73% of the queries (98,236) occurred only once. The most frequent query was *career services* with a count of 9587 and the next frequent query was *grades* with a count of 5727. On the list of queries that occurred 1000 times or more, all are education or campus life related: *tuition* (4837 queries), *housing* (4203), *transcripts* (2340), *transcript* (1211), *cheerleading* (1985), *cheerleaders* (1377), *registration* (1683), *football tickets* (1465), and *Webmail* (1317).

8.2.2   Trend Analysis of Web Searching

The search activities of an academic year are reflected in Figure 8.1. Here the number of queries per month shows similar patterns: the peak month is January for both years; June and August had comparatively fewer queries for both years. Further analysis of two selected weeks from Fall and Spring shows the Web activities across the days of a week (Figure 8.2).

Monday through Thursday generally had more queries than Friday and weekends. During the Christmas and New Year's holidays, the search activities dropped substantially, when the university had administrative closings. Such periods of reduced numbers of queries are certainly common for users who are members of the university community (students, faculty, and staff).

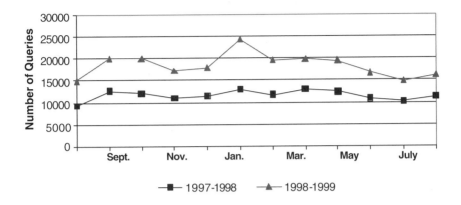

Figure 8.1. Monthly query totals for two academic years (1997–1998, 1998–1999).

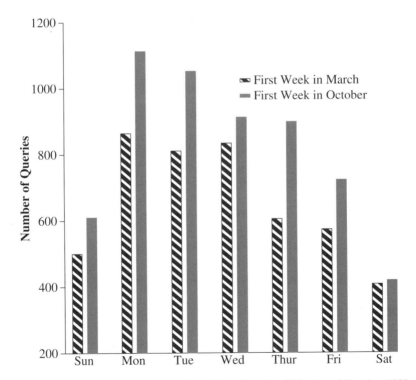

Figure 8.2. Number of queries per day for selected weeks of March and October 1999.

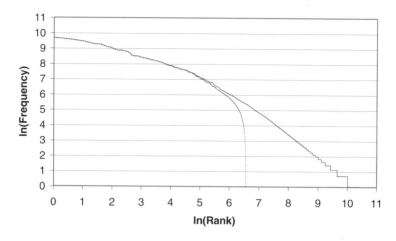

Figure 8.3. Natural log of term frequencies versus natural log of term ranks. Frequencies ranged from 1 to 16063 for occurrences ranging from 1 to 22829. Ranks for all terms (44652) and unique frequencies (705) are considered.

8.3   Zipf's Law

One of the enduring curiosities in text analysis and mining has been the fit of Zipf's law and its variations [BYRN99, Kor77]. Because query statements submitted to the search engines were from diverse users and extremely short as a text unit, it is difficult to predict if the vocabulary will show similar statistical distributions. Researchers of Excite query logs plotted lines of logarithmic rank-frequency distribution of terms and cleaned terms, resulting in lines far from (perfect) smooth straight lines of slope −1.

8.3.1   Natural Logarithm Transformations

Using the natural logarithm to transform the data collected for this study, two sets of data were plotted: one line representing ranks of all words by their frequencies in descending order, and one line representing ranks of unique frequencies (Figure 8.3). The former ignored the fact that many words had the same frequencies, especially in the low frequencies; the latter assumed fair ranking based on frequency and ignored the size of the vocabulary. Figure 8.3 illustrates how the two lines greatly overlap for ln(rank) up to 6 and then diverge by ln(rank) = 7. To understand if these lines show similar trends, four additional graphs were produced using the vocabulary of individual years (Figure 8.4). Figure 8.5 depicts two lines for word pairs, which show trends similar to those of single words. We show only 65535 term pairs (out of a total of 141353) for this figure, and note that only 49797 word pairs occurred in more than one query. In other words, 91556 (65%) word pairs share the frequency value 1. A comparison of all the curves shown indicates that the larger the vocabulary, the longer the tail. That is, a single straight trendline (linear fit) will not suffice to model query term (or term pair) frequency-ranking distributions.

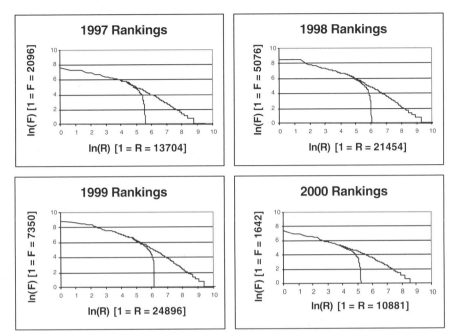

Figure 8.4. Term frequency rankings by year. The frequencies of all words are ranked first to generate the plot; the unique frequencies are also ranked (many terms have the same frequencies). Natural logarithms of term frequencies and ranks are plotted.

8.3.2   Piecewise Trendlines

The attempt to find the closest mathematical representation results in breaking the line at a point into two lines. This proves to be an effective way of modeling the trend. Figure 8.6 presents the trendline for all words using two equations: a line for the lower portion and a quadratic polynomial for the upper portion. For the lines representing unique frequencies, two lines will not be enough to achieve a close fit of the trendline because the low frequencies cluster many words (word pairs).

8.4   Vocabulary Growth

The more the number of queries in a year, the more the number of unique words or word pairs there will be. This change is not in proportion: the size of vocabulary increases much more slowly than the size of the queries (Figure 8.7). Further investigations into the overlapping of vocabulary across years revealed that 2912

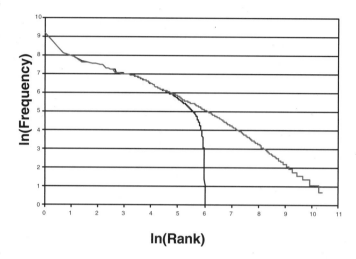

Figure 8.5. Frequency by rank for term pairs. The top curve reflects ranking of all occurrences of term pairs (65,535) and the bottom curve represents only unique frequencies of term pairs (411).

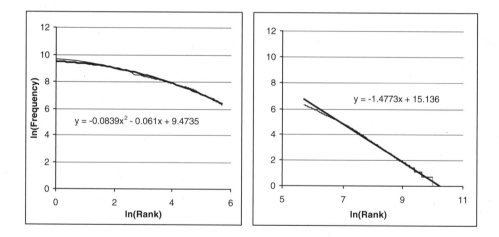

Figure 8.6. Quadratic and linear data fitting.

words occurred in all five years, 2534 words in four years, 3681 words in three years, 6547 words in two years, and 28168 words in one year. For all years, there was a total of 44652 unique words.

Only half of the words occurred in a total of 141353 pairs. There were 2222 pairs in all five years, 3617 pairs in four years, 7383 pairs in three years, 17586 pairs in two years, and 110545 pairs in only one of the years. Many of the high-frequency words or word pairs occurred in all years, such as of, and, services, student, career, and football are on top 10 list for all years; career-services is the top one pair for four years and third for 2001 when the log was incomplete; football tickets ranked 5th twice, 7th and top 13th once, but 1429th for 2001. Since the log file ended in May 2001 and football season starts in the Fall, some of the queries were certainly seasonal in nature.

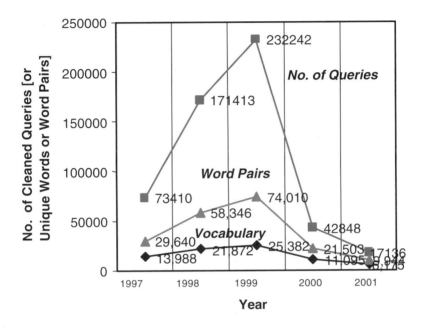

Figure 8.7. Trends in vocabulary growth.

8.5   Conclusions and Further Studies

This project revealed some characteristics of end-user searching on the Web, which can be exploited in search engine design. The occurrence of zero-hits due to the use of stop-words can be solved in two ways: the engine can automatically exclude them in query processing (many search engines now do this). This may not be enough, however, to improve searchers' skill through self-learning. An intelligent agent could present this fact at search time via a context-sensitive instruction.

Although the vocabulary does not fit Zipf's law closely, the frequency and ranking show regularities with simple polynomial data models underlying the use of words to form queries. The vocabulary size can be expected to increase slowly as the number of queries increases rapidly. The users of a specific website certainly share common interests. As reflected in this study, the users of an academic site, quite different from the users of general search engines, looked for information mostly related to education and campus life.

The analysis of word associations indicated that the matrix of vocabulary terms to term pairs is sparse with only a small portion of the vocabulary words co-occurring. Further explorations are needed to address questions such as: (1) How should a search engine interpret term pairs? (1a) Should the term pairs be treated as a Boolean AND operation? Or (1b) as a Boolean OR operation? (2) Should the words that always occur together be given higher weight even though their relative frequency (individually) is low? (3) For two-word queries, are both words good content descriptors or do they define a structure function? In other words, can one of the words form an effective query statement? To answer these questions, human judgments and additional tools are needed. The word associations measured by frequencies of word pairs may be predictable by word frequencies alone.

Web queries are short. Expansion of short queries by word associations may improve descriptions of information needs, which in turn increases precision. Techniques for clustering queries and generating categories of information needs are certainly needed. For the generation of content-based metadata, any system must be able to incorporate user vocabulary that can be automatically collected and analyzed.

Acknowledgments: This study was supported in part by a grant from the Computational and Information Sciences Interdisciplinary Council at the University of Tennessee, Knoxville.

References

[BYRN99]   R. Baeza-Yates and B. Ribeiro-Neto.*Modern Information Retrieval.*Addison-Wesley, Boston, 1999.

[JP01]   B.J. Jansen and U. Pooch.A review of Web searching studies and a framework for future research.*Journal of the American Society for Information Science and Technology*, 52(3):235–246, 2001.

[JSS00]   B.J. Jansen, A. Spink, and T. Saracevic.Real life, real users, and real needs: A study and analysis of user queries on the Web.*Information Processing and Management*, 36(2):207–227, 2000.

[Kor77]   R.R. Korfhage.*Information Storage and Retrieval.*Wiley, New York, 1977.

[RW00]   N. Ross and D. Wolfram.End user searching on the Internet: An analysis of term pair topics submitted to the Excite Search Engine.*Journal of the American Society for Information Science and Technology*, 51(10):949–958, 2000.

[SBC97]   B. Shneiderman, D. Byrd, and W.B. Croft.Clarifying search: A user-interface framework for text searches.*D-Lib Magazine*, 1:1–18, 1997.

[SHMM99]   C. Silverstein, M. Henzinger, H. Marais, and M. Moricz.Analysis of a very large Web search engine query log.*SIGIR Forum*, 33(1):6–12, 1999.

[SWJS01]   A. Spink, D. Wolfram, B. Jansen, and T. Saracevic.Searching the Web: The public and their queries.*Journal of the American Society for Information Science and Technology*, 52(3):226–234, 2001.

[Wol99]   D. Wolfram.Term co-occurrence in Internet search engine queries: An analysis of the Excite data set.*Canadian Journal of Information and Library Science*, 24(2/3):12–33, 1999.

[WP97]   P. Wand and L. Pouchard.End-user searching of Web resources: Problems and implications.In *Proceedings of the Eighth ASIS SIG/CR Workshop*, Washington DC, pages 73–85, 1997.

9

A Survey of Emerging Trend Detection in Textual Data Mining

April Kontostathis
Leon M. Galitsky
William M. Pottenger
Soma Roy
Daniel J. Phelps

Overview

In this chapter we describe several systems that detect emerging trends in textual data. Some of the systems are semiautomatic, requiring user input to begin processing, and others are fully automatic, producing output from the input corpus without guidance. For each Emerging Trend Detection (ETD) system we describe components including linguistic and statistical features, learning algorithms, training and test set generation, visualization, and evaluation. We also provide a brief overview of several commercial products with capabilities of detecting trends in textual data, followed by an industrial viewpoint describing the importance of trend detection tools, and an overview of how such tools are used.

This review of the literature indicates that much progress has been made toward automating the process of detecting emerging trends, but there is room for improvement. All of the projects we reviewed rely on a human domain expert to separate the emerging trends from noise in the system. Furthermore, we discovered that few projects have used formal evaluation methodologies to determine the effectiveness of the systems being created. Development and use of effective metrics for evaluation of ETD systems is critical.

Work continues on the semiautomatic and fully automatic systems we are developing at Lehigh University [HDD]. In addition to adding formal evaluation components to our systems, we are also researching methods for automatically developing training sets and for merging machine learning and visualization to develop more effective ETD applications.

9.1   Introduction

What is an emerging trend? An emerging trend is a topic area that is growing in interest and utility over time. For example, Extensible Markup Language (XML) emerged as a trend in the mid-1990s. Table 9.1 shows the results of an INSPEC® [INS] database search on the keyword "XML" from 1994 to 1999 (no records appeared before 1994). As can be seen from this table, XML emerged from 1994 to 1997; by 1998 it was well represented as a topic area.

Year	Number of Documents
1994	3
1995	1
1996	8
1997	10
1998	170
1999	371

Table 9.1. Emergence of XML in the Mid-1990s

Knowledge of emerging trends is particularly important to individuals and companies who are charged with monitoring a particular field or business. For example, a market analyst specializing in biotech companies might want to review technical and news-related literature for recent trends that will have an impact on the companies she is tracking. Manual review of all the available data is simply not feasible. Human experts who have the task of identifying emerging trends need to rely on automated systems as the amount of information available in digital form increases.

An Emerging Trend Detection application takes as input a collection of textual data and identifies topic areas that are either novel or are growing in importance within the corpus. Current applications in ETD fall generally into two categories: fully automatic and semiautomatic. The fully automatic systems take in a corpus and develop a list of emerging topics. A human reviewer then peruses these topics and the supporting evidence found by the system to determine which are truly emerging trends. These systems often include a visual component that allows the user to track the topic in an intuitive manner [DHJ+98, SA00]. Semiautomatic systems rely on user input as a first step in detecting an emerging trend [PD95, RGP02]. These systems then provide the user with evidence that indicates whether the input topic is truly emerging, usually in the form of user-friendly reports and screens that summarize the evidence available on the topic.

We begin with a detailed description of several semi- and fully automatic ETD systems in Section 9.2. We discuss the components of an ETD system including linguistic and statistical features, learning algorithms, training and test set generation, visualization, and evaluation. In Section 9.3 we review the ETD capabilities in commercial products. Our conclusions are presented in Section 9.4. In Section 9.5, Dr. Daniel J. Phelps, Leader of Eastman Kodak's Information Mining

Group, describes the role of ETD systems in modern corporate decision-making environments.

9.2   ETD Systems

As mentioned above, ETD systems can be classified as either fully automatic or semiautomatic. Semiautomatic systems require user input as a first step in detecting the emerging trends in a topic area. As part of our ongoing research at Lehigh University, we have developed both fully and semiautomatic systems that have successfully identified emerging trends. In this section we provide an overview of the components that are included in most ETD systems (input data sets, attributes used for processing, learning algorithms, visualization, evaluation), followed by a detailed description of several ETD systems.

We begin with a discussion of the data that are used in ETD systems. The most commonly used data repository for ETD emerged from the Topic Detection and Tracking (TDT) project [TDT] that began in 1997. TDT research develops algorithms for discovering and threading together topically related material in streams of data, such as newswire and broadcast news, in both English and Mandarin Chinese. The TDT project, while not directly focused on emerging trend detection, has nonetheless encouraged the development of various fully automated systems that track topic changes through time. Several of those algorithms are described in this section.

As part of the TDT initiative several data sets have been created. The TDT data sets are sets of news stories and event descriptors. Each story/event pair is assigned a relevance judgment. A relevance judgment is an indicator of the relevance of the given story to an event. Table 9.2 portrays several examples of the relevance judgment assignment to a story/event pair. Thus the TDT data sets can be used as both training and test sets for ETD algorithms. The Linguistic Data Consortium (LDC) [Lin] currently has three TDT corpora available for system development: the TDT Pilot study (TDT-Pilot), the TDT Phase 2 (TDT2), the TDT Phase 3 (TDT3), as well as the TDT3 Arabic supplement.

Not all of the systems we describe rely on the TDT data sets. Other approaches to the creation of test data have been used, such as manually assigning relevance judgments to the input data and comparing the system results to the results produced by a human reviewer. This approach is tedious and necessarily limits the size of the data set. Some of the systems we present use databases such as INSPEC®, which contains engineering abstracts, or the United States patent database [US], which allows searching of all published US patents. The input data set, along with the selection of appropriate attributes that describe the input, is a critical component of an ETD system. Attribute selection is at the core of the tracking process, since it is the attributes that describe each input item and ultimately determine the trends.

The attributes obtained from the corpus data are input to the methods/techniques employed by each ETD system we describe below. As shown, some research groups

Story Description	Event	Relevance Judgment
Story describes survivor's reaction after Oklahoma City Bombing	Oklahoma City Bombing	Yes
Story describes survivor's reaction after Oklahoma City Bombing	US Terrorism Response	No
Story describes FBI's increased use of surveillance in government buildings as a result of the Oklahoma City Bombing	Oklahoma City Bombing	Yes
Story describes FBI's increased use of surveillance in government buildings as a result of the Oklahoma City Bombing	US Terrorism Response	Yes

Table 9.2. Story/Event pairs

use traditional Information Retrieval (IR) methodologies to detect emerging trends, while others have focused more on traditional machine learning approaches such as those used in data mining applications.

Work in the areas of visualization-supported trend detection has explored multiple techniques for identifying topics. When a user is trying to understand a large amount of data, a system that allows an overview, at multiple levels of detail and from multiple perspectives, is particularly helpful. One of the simplest approaches is a histogram, where bars indicate discrete values of actual data at some discrete point in time. Information visualization is meant to complement machine learning approaches for trend detection. Plotting the patterns along a timeline allows one to see the rate of change of a pattern over time. For each algorithm described below, we discuss the visualization component, showing how the component enhances the trend detection capabilities of the system.

The evaluation of an emerging trend detection system can be based on formal metrics, such as precision (the percentage of selected items that the system got right) and recall (the proportion of the target items that the system found), or by less formal, subjective means (e.g., answers to usability questions such as: Is the visualization understandable?). The particulars of an evaluation are related to the goals of the method and thus can vary greatly, but some justification and interpretation of the results should always exist to validate a given system.

9.2.1 Technology Opportunities Analysis (TOA)

Alan L. Porter and Michael J. Detampel describe a semiautomatic trend detection system for technology opportunities analysis in [PD95]. The first step of the process is the extraction of documents (such as INSPEC® abstracts) from the knowledge area to be studied. The extraction process requires the development of a list of potential keywords by a domain expert. These keywords are then combined into queries using appropriate Boolean operators to generate comprehensive and accurate searches. The target databases are also identified in this phase (e.g., INSPEC®, COMPENDEX® [COM], US Patents, etc.).

The queries are then input to the Technology Opportunities Analysis Knowbot (TOAK), a custom software package also referred to as TOAS (Technology Opportunities Analysis System). TOAK extracts the relevant documents (abstracts) and provides bibliometric analysis of the data. Bibliometrics uses information such as word counts, date information, word co-occurrence information, citation information, and publication information to track activity in a subject area. TOAK facilitates the analysis of the data available within the documents. For example, lists of frequently occurring keywords can be quickly generated, as can lists of author affiliations, countries, or states.

In [PD95], the authors present an example of how the TOAK system can be used to track trends in the multichip module subfield of electronic manufacturing and assembly. Figure 9.1 [PD95] shows a list of keywords that appear frequently with "multichip module" in the INSPEC® database. The authors observed that multichip modules and integrated circuits (particularly hybrid integrated circuits) co-occurred very frequently. An additional search using the US Patent database showed that many patents had been issued in the area of multichip modules. Furthermore, the integrated circuits activity was more likely to be US-based, while large-scale integration activity was more likely to be based in Japan.

TOAK is meant to be used by a human expert in an interactive and iterative fashion. The user generates initial queries, reviews the results and is able to revise the searches based on his domain knowledge. TOA represents an alternative approach to the time-consuming literature search and review tasks necessary for market analysis, technology planning, strategic planning, or research [PD95].

Input Data and Attributes

The INSPEC® database serves as the primary corpus for TOA and its related software, TOAK. Two opportunities exist for attribute selection. First (Table 9.3), a list of keywords (a single word or multiple words, termed n-grams[1]) and their possible combinations (using Boolean operators) are supplied to TOAK, which retrieves all relevant items. The number of keyword occurrences and keyword co-occurrences (the appearance of two keywords in the same item) are calculated per

---

[1] An n-gram is a sequence of n words. For example, the phrase "stock market" is a bigram (or 2-gram).

Multichip Module Keywords and Frequencies
[INSPEC Database]

Keyword	Number of articles	Keyword	Number of articles
Multichip modules	842	Circuit layout CAD	69
Packaging	480	Tape automated bonding	68
Hybrid integrated circuits	317	Printed circuit manufacture	66
Module	271	Printed circuit design	65
Integrated circuit technology	248	Thin film circuit	62
Integrated circuit testing	127	CMOS integrated circuits	56
Substrates	101	Soldering	50
VLSI	98	Optical interconnections	48
Surface mount technology	93	Lead bonding	44
Flip-chip devices	93	Integrated optoelectronics	43
Integrated circuit manufacture	88	Printed circuits	42
Ceramics	85	Production testing	41
Circuit reliability	80	Reliability	41
Polymer films	79	Microassembling	38
Cooling	70	Circuit CAD	35
Metallisation	69	Microprocessor chips	35

Figure 9.1. Co-occurrences with "multichip modules" [PD95]

year and over all years. A second pass (Table 9.4) involves selecting all phrases (single- and multiword) from a specific field and calculating the number of items that contain each phrase. For example, every phrase in the keyword field of each item may be counted, or each phrase in the affiliation field [PD95].

Attribute	Detail	Generation
n-grams	E.g., multichip modules, ball grid array	Manual
Frequency	Count of *n*-gram occurrence	Automatic
Frequency	Count of *n*-gram co-occurrence	Automatic
Date	Given by year	Automatic

Table 9.3. TOA First Pass Attributes

Attribute	Detail	Generation
Field	A section of an item (e.g., an indexing term or city name)	Manual
Frequency	Count of *n*-gram occurrence in a field	Automatic

Table 9.4. TOA Second Pass Attributes

Learning Algorithms

Like most of the systems that facilitate trend detection in textual collections, TOA relies on the expertise of the user who is researching a given area. TOAK provides access to many different data sources, including INSPEC®, COMPENDEX®, US Patents, and others, but is necessarily limited as not all R&D work is patented or published. The power of TOAK resides in the visual interface and easy access to different views of the data. There are no inherent learning algorithms present in the system; the user is solely responsible for trend detection.

Visualization

Visualizations in TOA include frequency tables, histograms, weighted ratios, log–log graphs, Fisher–Pry curves, and technology maps [PD95]. These tools present information graphically using various linking and clustering approaches such as multidimensional scaling. TOA can generate visualizations based on attributes such as keyword, organization, document source, country of origin, and author. Principal components visualizations that represent relationships among conceptual clusters can also be created.

Evaluation

Identification of trends is left to the user in this semiautomatic method. TOA could, however, be evaluated on how well it presents information to the user. Visualizations are meant to significantly increase understanding of the data, and intuitively do. TOA developers, however, provide no evidence for the efficacy of these tools, apart from various author's claims. Solutions do however exist for evaluating this type of method. For example, results of usability studies and focus groups can strengthen arguments that visualization is indeed helpful. ThemeRiver™ (Section 9.2.5) employs this usability approach for evaluation. Formal metrics, even with a semiautomatic method, can also be utilized as in CIMEL (Section 9.2.2).

9.2.2   CIMEL: Constructive, Collaborative Inquiry-Based Multimedia E-Learning

CIMEL is a multimedia framework for constructive and collaborative inquiry-based learning that we, the authors of this survey, have developed [BPK$^+$01, BPK$^+$02, CIM]. Our semiautomatic trend detection methodology described in [RGP02] has been integrated into the CIMEL system in order to enhance computer science education. A multimedia tutorial has been developed to guide students through the process of emerging trend detection. Through the detection of incipient emerging trends, students see the role that current topics play in course-related research areas. Early studies of this methodology, using students in an upper-level computer science course, show that use of the methodology improves the number of incipient emerging trends identified.

   Our semiautomatic algorithm employs a more robust methodology than TOA because the user base is assumed to be individuals who are learning a particular area, as opposed to domain experts. The methodology relies on web resources to identify candidate emerging trends. Classroom knowledge, along with automated "assistants," help students to evaluate the identified candidate trends. This methodology is particularly focused on incipient trends.

1. Identify a main topic area for research (e.g., object databases)
2. Identify recent conferences and workshops in this area (e.g., OOPSLA for object-oriented programming)
3. Review content and create a list of candidate emerging trends
4. Evaluate each emerging trend identified in Step 3, using general web research tools (e.g., Google™ search)
5. For each candidate emerging trend remaining after Step 4, verify the trend using an INSPEC® database search

Table 9.5. Methodology for Detecting Emerging Trends

The methodology is outlined in Table 9.5. In Step two of this methodology (after a main topic area has been identified) the user is directed to recent conferences and workshops online and instructed to review the content and develop a list of candidate emerging trends. Next, the user is directed to a general-purpose web search engine to find other references to candidate emerging trends identified in step three. Searches using the candidate trend phrase, along with terms such as "recent research," "approach," and so on, are employed to improve the precision of the search results. The user is provided with a detailed algorithm that includes parameters for evaluation of the pages returned from the search engine. The candidate emerging trend may be rejected as a result of this search. In addition, other candidate emerging trends may be identified in this step.

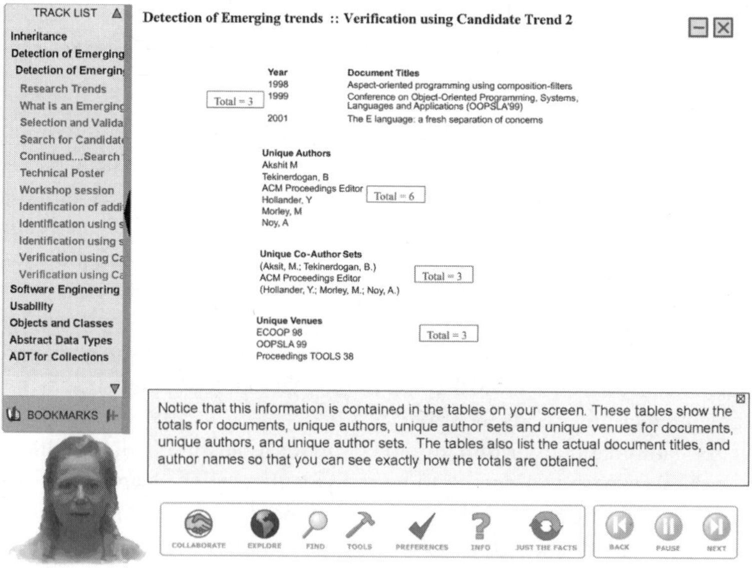

Figure 9.2. Emerging trend detection tutorial [CIM].

Finally, the user is asked to verify candidate emerging trends using document count and author and publication venue spread based on an INSPEC® database search. To make the trend detection process easier, this step has been automated [Gev02]. The user need only enter a candidate emerging trend (Figure 9.2) identified in Steps three and/or four, and the database search tool automatically generates document count, unique author sets, unique coauthor sets, and a list of unique venues (all across time) that pertain to the chosen candidate emerging trend. The tool also provides a link to the corresponding abstracts, which can be accessed by clicking on individual document titles. This feature of the tool is important, as the user still has to decide whether a given candidate trend is truly emerging based on heuristics provided in the tutorial.

For example, students in an upper-level object-oriented software engineering course might be asked to find an emerging trend in the field of object databases. Several conference websites would be provided, including the Conference on Object-Oriented Programming, Systems, Languages, and Applications (OOPSLA) website. A manual review of the content of papers presented at OOPSLA '01 leads the student to the candidate emerging trend "XML Databases." A search of the Web using Google™ results in additional papers related to XML databases, providing further evidence that "XML Databases" is an emerging trend. Finally, the student is directed to the INSPEC® database. A search using XML and Databases and Object-oriented reveals the information depicted in Table 9.6. Further inspection reveals multiple author sets and publication venues, confirming that "XML Databases" is an incipient emerging trend in the field of object databases.

Year	Number of Documents
Prior to 1999	0
1999	5
2000	11
2001	5

Table 9.6. "XML Databases" Is an Emerging Trend [RGP02]

Input Data and Attributes

The corpus for this semiautomatic methodology can be any web resource. A description of the main topic is chosen, which can consist of any text. An initial search of recent conferences and workshops is performed to identify candidate emerging trends. Using a web search engine, phrases associated with emerging trends[2] are used in conjunction with either the main topic or the candidate emerging trends to uncover additional evidence for the given candidate and/or to identify

---

[2]"Supporting" terms – the list of current associated "supporting" terms: most recent contribution, recent research, a new paradigm, hot topics, emergent, newest entry, cutting-edge strategies, first public review, future, recent trend, next generation, novel, new approach, proposed, and current issues.

other candidate trends. Several attributes guide this initial decision-making process (Table 9.7), including the current year, the number of times either the main topic or candidate emerging trend appears on the page, the number of supporting terms on the page, and the line or paragraph containing the main topic/candidate emerging trend and supporting term [RGP02]. The validation step (Table 9.8) involves automatically calculating four frequencies across time: the number of unique documents, unique authors, unique author sets, and unique venues [Gev02]. These frequencies help the user make a final emerging trend determination. For example, an increase in the number of documents that reference the main topic and candidate emerging trend over time is indicative of a true emerging trend. On the other hand, if one or two documents appear in different years by the same author, the candidate trend may not actually be emerging [RGP02].

Attribute	Detail	Generation
n-grams	Main topic, e.g., "object databases"	Manual
n-grams	Candidate trend, e.g., "XML databases"	Manual
n-grams	Supporting terms	Automatic
n-grams	Search item – any Boolean <and> combination of the previous attributes, e.g., "XML <and> novel"	Automatic
Date	Given by year	Automatic
Frequency	Count of main topic/candidate trend in page	Automatic
Frequency	Count of "supporting" terms	Automatic
n-grams	Line or paragraph containing the main topic/candidate trend and "supporting" term in a given document	Manual

Table 9.7. CIMEL Initial Step Attributes

Attribute	Detail	Generation
Frequency	Count of unique authors, per year	Automatic
Frequency	Count of unique documents, per year	Automatic
Frequency	Count of unique author sets, per year	Automatic
Frequency	Count of unique venues, per year	Automatic

Table 9.8. CIMEL Validation Step Attributes

Learning Algorithms

Like TOA, the CIMEL system relies on the user to detect emerging trends. No machine learning component is employed. Instead CIMEL relies on a precisely defined manual process. Like TOA, this system is restricted by the electronic availability of documentation in a given subject area. Furthermore, the INSPEC® query tool is currently based on abstracts that are downloaded to a local database, which must be periodically refreshed. Unlike TOA, CIMEL provides specific parameters for identifying an emerging trend, rather than relying solely on the domain expertise of the user.

Visualization

At the current time the visualization component for trend detection in CIMEL is under development.

Evaluation

Several experiments have been conducted to evaluate the utility of the ETD component of the CIMEL system. In one such experiment, two groups of students in a programming languages class were asked to identify emerging trends in the area of inheritance in object-oriented programming. Group B (experimental) viewed a multimedia tutorial on the methodology that included a case study; Group A (control) did not. Hypothesis testing was performed using the standard metric of precision. Precision for a student was calculated by dividing the number of actual emerging trends found (zero, one or two for this experiment) by the number of total trends found (two, if the student completed the assignment successfully). Recall was not determined since a complete list of emerging trends was not available. A lower tail t-test concluded with 95% confidence that the mean precision of students that used the methodology (Group B) was significantly greater than the mean precision of students that did not use the methodology (Group A). These results provide convincing evidence that the ETD methodology employed in the CIMEL system is effective at detecting emerging trends [Roy02].

9.2.3   TimeMines

The TimeMines system [SJ00] takes free text data, with explicit date tags, and develops an overview timeline of statistically significant topics covered by the corpus. Figure 9.3 presents sample output from TimeMines. TimeMines relies on Information Extraction (IE) and Natural Language Processing (NLP) techniques to gather the data. The system employs hypothesis-testing techniques to determine the most relevant topics in a given timeframe. Only the "most significant and important" information (as determined by the program) is presented to the user.

TimeMines begins processing with a default model that assumes the distribution of a feature depends only on a base rate of occurrence that does not vary with time. Each feature in a document is compared to the default model. A statistical test is used to determine if the feature being tested is significantly different than what the model would expect. If so, the feature is kept for future processing; otherwise it is ignored.

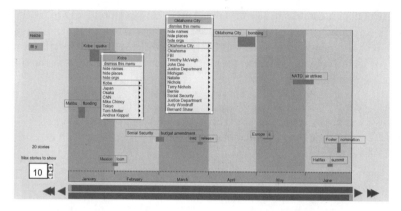

Figure 9.3. TimeMines Sample Output [SJ00].

The reduced set of features that is developed using the first round of hypothesis testing is then input into a second processing phase which groups related features together. The grouping again relies on probabilistic techniques that combine terms that tend to appear in the same timeframes into a single topic. Finally, a threshold is used to determine which topics are most important and these are displayed via the timeline interface (Figure 9.3). The threshold is set manually, and is determined empirically.

Like TOA, TimeMines presents a model of the data without drawing any specific conclusions about whether a topic is emergent. It simply presents the most statistically significant topics to the user, and relies on the user's domain knowledge for evaluation of the topics.

Input Data and Attributes

In [SJ00], the TDT and TDT-2 corpora were date tagged and part-of-speech tagged with JTAG [XBC94]. (TDT-2 was preliminarily tagged with Nymble [BMSW97].) In the TimeMines system, an initial attribute list of all "named entities" and certain noun phrases is generated. A named entity is defined as a specified person, location, or organization (extracted using the Badger IE system [FSM+95]). Noun phrases match the regular expression (N|J)* N for up to five words, where N is a noun, J is an adjective, | indicates union, and * indicates zero or more occurrences. The documents are thus represented as a "bag of attributes," where each attribute is true or false (i.e., whether the named entity or noun phrase is contained in the document). The attributes are shown in Table 9.9.

Attributes	Detail	Generation
Named Entity	Person, location, or organization	Automatic
n-grams	Follows (N|J)*N pattern for up to five words, e.g., "textual data mining"	Automatic
Presence	"True" if the named entity or *n*-gram occurs in the document, else "False." Each document has a presence attribute for each named entity and *n*-gram	Automatic
Date	Given by day	Automatic

Table 9.9. TimeMines Attributes

Learning Algorithms

There are two separate machine learning aspects present in the TimeMines application. First, as stated before, TimeMines must select the "most significant and important information" to display. To do this, TimeMines must extract the "most significant" features from the input documents.

TimeMines uses a statistical model based on hypothesis testing to choose the most relevant features. As noted, the system assumes a stationary random model for all features (*n*-grams and named entities) extracted from the corpus. The stationary random model assumes that all features are stationary (meaning their distributions do not vary over time) and the random processes generating any pair of features are independent. Features whose actual distribution matches this model are considered to contain no new information and are discarded. Features that vary greatly from the model are kept for further processing. The hypothesis testing is time dependent. In other words, for a specific block of time, a feature either matches the model (at a given threshold) or violates the model. Thus the phrase "Oklahoma City Bombing" may be significant for one time slice, but not significant for another.

After the feature set has been pruned in this manner, TimeMines uses another learning algorithm, again based on hypothesis testing. Using the reduced feature set, TimeMines checks for features within a given time period that have similar distributions. These features are grouped into a single *topic*. Thus each time period may be assigned a small number of topic areas, represented by a larger number of features.

One potential drawback of ranking the general topics derived from the significant attributes is discussed in [SJ00]. The occurrence of an attribute is measured against all other occurrences of it in the corpus. As a result a consistently heavily used attribute may not distinguish itself properly. The Kenneth Starr–President Clinton investigation is unquestionably the most covered story in the TDT-2 corpus, yet ranked twelfth because it is so prevalent throughout. Against a longer time period, including time after coverage had died down, the story probably would have ranked first.

Like all of the algorithms we present here, the final determination of whether a topic is emerging is left to the user, but unlike CIMEL and TOA, the user does not direct the TimeMines system. This system is completely automated; given a time-tagged corpus it responds with a graphical representation of the topics that dominate the corpus during specific time periods.

Visualization

TimeMines generates timelines automatically for both visualization of temporal locality of topics and the identification of new information within a topic. The x-axis represents time, while the y-axis represents the relative importance of a topic. The most statistically significant topic appears near the top of the visualization (Figure 9.3). Each block in the visualization interface includes all the terms used to describe a topic and thus indicates the coverage within the corpus. Clicking on a term (named entity or n-gram) pops up a menu of all the associated features of that type within the topic, and a submenu option allows the user to choose this feature as the label, or to obtain more information about the feature. However, no effort is made to infer any hierarchical structure in the appearance of the feature in the timeline.

Evaluation

Two hypotheses are evaluated in [SJ00]: do term occurrence and co-occurrence measures properly group documents into logical time-dependent stories, and, are the stories themselves meaningful to people? A randomization test [Edg95] was conducted to support the first hypothesis. The documents were shuffled and assigned an alternate date, but were otherwise left intact. From an IR standpoint the corpus looked the same, since term frequency and inverse document frequency were preserved. The authors concluded that the results of this test overwhelmingly suggest the groupings are logical and not random.

The second hypothesis was explored with two methods of evaluation but results were inconclusive. The first evaluation method used precision and recall metrics from IR. The January 1996 *Facts on File* [Fac] listed 25 major stories, which were used as the "truth" set to compare with the TimeMines-generated major stories. Recall was defined as the number of *Facts on File* major stories identified by TimeMines divided by the total number of *Facts on File* major stories. Precision was defined as the number of *Facts on File* major stories identified by TimeMines divided by the total number of TimeMines-identified major stories. A relatively low precision of 0.25 and a similarly low recall of 0.29 resulted.

The second evaluation method attempted to tune the threshold. Four students manually determined whether the automatic groupings related to zero, one, or multiple topics. Based on a pairwise Kappa statistic, however, the manual results could not be distinguished from random results [SA00].

9.2.4   New Event Detection

New event detection, also referred to as first story detection, is specifically included as a subtask in the TDT initiative. New event detection requires identifying those news stories that discuss an event that has not already been reported in earlier stories. New event detection operates without a predefined query. Typically algorithms look for keywords in a news story and compare the story with earlier stories. The approach taken in [APL98] implies that the input be processed sequentially in date order: that is, only past stories can be used for evaluation, not the entire corpus.

A new event detection algorithm based on a single-pass clustering algorithm is presented in [APL98]. The content of each story is represented as a query. When a new story is processed, all the existing queries (previous stories) are run against it. If the "match" exceeds a predefined threshold (discussed below) the new story is assumed to be a continuation of the query story. Otherwise it is marked as a new story.

An interesting characteristic of news stories is that events often occur in bursts. Figure 9.4 [YPC98] portrays a temporal histogram of an event where the x-axis represents time in terms of days (1 through 365) and the y-axis is the story count per day.

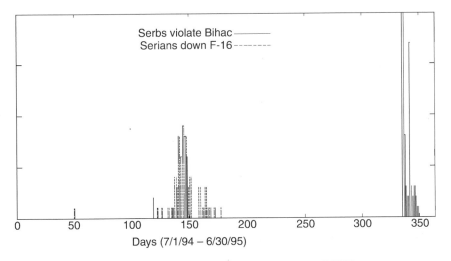

Figure 9.4. Temporal histogram of news data [YPC98].

News stories discussing the same event tend to be in temporal proximity and hence lexical similarity and temporal proximity are considered to be two criteria for document clustering. Also, a time gap between the bursts as exemplified in Figure 9.4 discriminates between distinct events, and the system is more likely to match stories that appear in the same timeframe.

As reported in [APL98], with proper tuning the algorithm was able to separate news stories related to the Oklahoma City Bombing from those about the World

Trade Center bombing. However, some stories could not be detected. For example, the crash of Flight 427 could not be distinguished from other airplane accidents, and the O.J. Simpson trial could not be separated from other court cases.

Input Data and Attributes

All stories in the TDT corpus deemed relevant to 25 selected "events" were processed. For new event detection, each story was represented by a query and threshold. Table 9.10 lists all the attributes required for computing the query and threshold. The n most frequent single words comprise the query, and are weighted and assigned a "belief" value by the Inquery system [ABC+95], indicating the relevance of each word in the story to the query. Belief is calculated using term frequency and inverse document frequency. Term frequency is derived from the count of times the word occurs in the story, the length of the story, and the average length of a story in the collection. Inverse document frequency is derived from the count of stories in the collection and the count of stories that contain the word.

Attribute	Detail	Generation
Unigram	A single word	Automatic
Frequency	Number of times unigram occurs, per story	Automatic
Count	Total number of unigrams, per story	Automatic
Mean	Average number of unigrams per story	Automatic
Frequency	Number of stories in which unigram occurs	Automatic
Count	Number of stories	Automatic
Date	Given by available granularities	Automatic

Table 9.10. New Event Detection Attributes

Learning Algorithms

As noted, the approach presented in [APL98] is based on a single-pass clustering algorithm that detects new stories by comparing each story processed to all of the previous stories/queries detected. As each incoming story is processed, all previous "queries" are run against it. If a story does not match any of the existing queries, the story is considered a new event.

The system relies on a threshold to match the queries to the incoming stories. The initial threshold for a query is set by evaluating the query with the story from which it originated. If a subsequent story meets or exceeds this initial threshold for the query, the story is considered a match. The threshold is used as input to a thresholding function based on the Inquery system described above [ABC+95]. Since new event detection implies that documents are processed in order, however, traditional IR metrics that are usually applied to an entire corpus (such as the number of documents containing the term and average document length) are not readily available. To overcome this problem, an auxiliary collection is used to provide this information to the Inquery system. The thresholding function takes advantage of the time dependent nature of the news story collection by using a time penalty that increases the value required to "match" a story as stories grow farther apart in time.

Like the TimeMines system, the new event detection system described here is completely automated. Given a corpus, it provides a list of "new events" in the form of news stories that first describe an occurrence of an event. New event detection differs somewhat from ETD in that it is focused on the sudden appearance of an unforeseen event rather than the (more gradual) emergence of a trend.

Visualization

The new event detection system is based on Lighthouse [LA00b, LA00a], an interactive information retrieval system that provides a ranked list of search results together with two- and three-dimensional visualizations of interdocument similarities. After events are extracted, a visual timeline is constructed to show how these events occur in time and relate to each other.

Evaluation

[APL98] evaluated their system using miss (false negative) and false alarm (false positive) rates as well as the metrics of precision and recall. Arriving at meaningful thresholds for these rates was difficult, and as a complement, Detection Error Trade-off (DET) curves [MDOP97] were studied. DET curves highlight how miss and false alarm rates vary with respect to each other (each is plotted on an axis in a plane). A perfect system with zero misses and false alarms would be positioned at the origin, thus, DET curves "closer" to the origin are generally better. *Close* was defined as the Euclidean distance from the DET curve to the origin in [APL98]. Using nearly all (400) single-word attributes in the queries resulted in averages of 46% for the miss rate, 1.46% for the false alarm rate, 54% for recall, and 45% for precision.

9.2.5   ThemeRiver™

Similar to TimeMines, ThemeRiver™ [HHWN02] summarizes the main topics in a corpus and presents a summary of the importance of each topic via a graphical user interface. The topical changes over time are shown as a *river* of information. The river is made up of multiple streams. Each stream represents a topic and each topic is represented by a color and maintains its place in the river relative to other topics. Figure 9.5 portrays an example visualization.

The river metaphor allows the user to track the importance of a topic over time (represented on the horizontal axis). The data represented in Figure 9.5 are from Fidel Castro's speeches. You can see that Castro frequently mentioned oil just before American oil refineries were confiscated in 1960 (shown as the second vertical

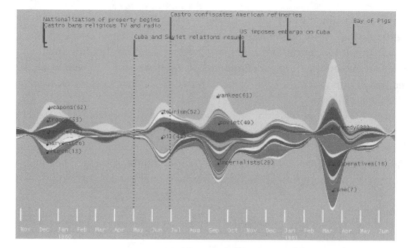

Figure 9.5. ThemeRiver™ sample output [HHWN02].

line from the left in Figure 9.5). Oil is the large bubble immediately preceding this dotted line in the middle of the river. At no other time did Castro dwell on that topic in the 18-month period represented by this corpus.

Such patterns in the data may confirm or refute the user's knowledge of hypotheses about the collection. Like TOA and TimeMines, ThemeRiver™ does not presume to indicate which topics are emergent. The visualization is intended to provide the user with information about the corpus. ThemeRiver™ presents a topic- or feature-centered view of the data. This topic-centered view is a distinguishing characteristic of the ETD approaches surveyed in this chapter. Related areas in information retrieval, such as text filtering and text categorization, are usually document-centered.

Input Data and Attributes

The corpus in the example presented in [HHWN02] consisted of speeches, interviews, articles, and other text about Fidel Castro over a 40-year period. ThemeRiver™ automatically generates a list of possible topics, called *theme words*, of which a subset is manually chosen as attributes (the example in [HHWN02] narrowed the list to 64). Counts of the number of documents containing a particular theme word for each time interval provide the input for the method. An alternate count, using the number of occurrences of the theme word for each time interval is suggested but not implemented in [HHWN02].

An automatic method for generating the initial list of theme words was not specified, nor was the procedure for deciding which or how many of the theme words should be included in the subset. Theme word frequencies are computed after these attributes are chosen, effectively making attribute selection a manual process (i.e., not automatic based strictly on the counts; see Table 9.11).

Attribute	Detail	Generation
Unigram	A single word	Manual
Frequency	Number of documents in which unigram occurs, per time interval	Automatic
Date	Given by month	Automatic

Table 9.11. ThemeRiver™ Attributes

Learning Algorithms

The ThemeRiver™ application does not use a learning algorithm per se. Like TOA, it provides a view of the data that an experienced domain expert can use to confirm or refute a hypothesis about the data. ThemeRiver™ begins by binning time-tagged data into time intervals. A set of terms, or themes, that represent the data is chosen and the river is developed based on the strength of each theme in the collection. As noted, the themes are chosen by automatically developing a list of words that are present in the data and then manually selecting a subset that represents various topics. The number of documents containing the word determines the strength of each theme in each time interval. Other methods of developing the themes and strengths are possible. The visual component of ThemeRiver™ is the most important aspect of this work, particularly as it applies to trend detection.

Visualization

The ThemeRiver™ system uses the river metaphor to show the flow of data over time (Figure 9.5). While the river flows horizontally, vertical sections of the river contain colored currents that identify topics or themes. The width of the river changes with the emergence or disappearance of topics, thereby making the system effective in cases where there is no major variation in topic.

The curves in Figure 9.5 show how interpolation is done to obtain a river metaphor. The idea is to produce a smooth curve with positive stream width for better visual tracking of the stream across time. Even though this technique aids human pattern recognition, a simple histogram can be more accurate. The algorithm interpolates between points to generate smooth curves (continuity in the flow of the river).

ThemeRiver™ makes judicious use of color, leveraging human perceptual and cognitive abilities. Themes are sorted into related groups, represented by a color family. This allows viewing of a large number of (related) themes that can easily be separated due to color variation. For example, "germany", "unification", "gdr", and "kohl" can be represented by different shades of the same color and hence can easily be identified as being related.

Evaluation

Evaluation, or usability in such visual applications, was conducted with two users in [HHWN02]. After being given some background information about the data, the users were asked about specifics related to the following five general questions.

- Did the users understand the visualization?

- Could they determine differences in theme discussion?

- Did the visualization prompt new observations about the data?

- Did the users interpret the visualization in any unexpected ways?

- How did the interpretation of the visualization differ from that of a histogram?

Observation, verbal protocol, and a questionnaire were used to gather feedback. This evaluation method is formalized well, but it lacks significance due to the small sample consisting of just two users.

9.2.6   PatentMiner

The PatentMiner system was developed to discover trends in patent data using a dynamically generated SQL query based upon selection criteria input by the user [LAS97]. The system is connected to an IBM DB2 database containing all granted United States (US) patents. There are two major components to the system, phrase identification using sequential pattern mining [AS95, SA96] and trend detection using shape queries.

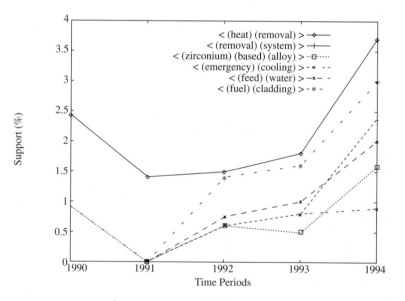

Figure 9.6. PatentMiner sample output [LAS97].

Input Data and Attributes

As noted, in [LAS97] an IBM DB2 database containing all US Patents served as the basis for the corpus. Several procedures prepare the data for attribute extraction. Stop-words are removed. Identifiers are assigned to the remaining words, indicating position in the document and occurrences of sentence, paragraph, and section boundaries. After a subset of patents is specified by category and date range, the Generalized Sequential Patterns (GSP) algorithm [SA96] selects user-defined attributes, called *phrases*. Only phrases with support greater than a user-defined minimum are considered. A phrase can be any sequence of words, with a minimum and maximum *gap* between any of the words. Gaps can be described in terms of words, sentences, paragraphs, or sections. For example, if the minimum sentence gap is one for the phrase "emerging trends", than "emerging" and "trends" must occur in separate sentences. Or if the maximum paragraph gap is one, than "emerging" and "trends" must occur in the same paragraph. A time window indicates the number of words to group together before counting gap length. Finally, a shape definition language (SDL) [APWZ95] specifies which types of trends (e.g., upwards, spiked, etc.) are displayed. Table 9.12 summarizes these attributes.

The number of phrases selected can be substantial, given their very open-ended nature. Two pruning methods are discussed in [LAS97]. A subphrase of a phrase may be ignored if the support of the two phrases is similar. Or, a subphrase (general, higher-level) might be preferred over a longer phrase (specific, lower-level) initially, after which specific lower-level phrases could be easier to identify. This has the flavor of the technique used in CIMEL in which a main topic is combined with a candidate trend in order to improve the precision of the results.

Attribute	Detail	Generation
n-grams	Search phrase, e.g., emerging trends	Manual
Size	Minimum gap, with distinct gaps for words, sentences, paragraphs, and sections	Manual
Size	Maximum gap, with distinct gaps for words, sentences, paragraphs, and sections	Manual
Size	Time window, groups words in a phrase before determining gaps	Manual
Ratio	Support, number of search phrases returned divided by total number of phrases	Manual
Date	Given by available granularities	Manual
Shape	Graphical trend appearance over time, e.g., spiked or downwards	Manual

Table 9.12. PatentMiner Attributes

Learning Algorithms

Most of the systems presented in this survey use traditional IR techniques to extract features from the text corpus that serves as input; the PatentMiner system takes a different approach. PatentMiner adapts a sequential pattern matching technique that is frequently used in data mining systems. This technique treats each word in the corpus as a transaction. The pattern matching system looks for frequently occurring patterns of words. The words may be adjacent, or separated by a variable number of other words (up to some maximum that is set by the user). This technique allows the system to identify frequently co-occurring terms and treat them as a single topic. [LAS97] refers to the resulting set of words (that make up a topic) as a "phrase."

As with TimeMines, documents in the input data set are binned into various collections based on their date information. The above technique is used to extract phrases from each bin and the frequency of occurrence of each phrase in all bins is calculated. A shape query is used to determine which phrases to extract, based on the user's inquiry.

The shape query processing is another learning tool borrowed from data mining [APWZ95]. In the PatentMiner system, the phrase frequency counts represent a data store that can be mined using the shape query tool. The shape query has the ability to match upward and downward slopes based on frequency counts. For example, a rapidly emerging phrase may occur frequently in two contiguous time slices, then level off, before continuing on an upward trend. The shape query allows the user to graphically define various shapes for trend detection (or other applications) and retrieves the phrases with frequency distributions that match the query.

Like ThemeRiver™, TimeMines, and others, the PatentMiner system presents a list of phrases to the user. The domain expert must then identify those that represent emerging trends.

Visualization

The system is interactive; a histogram is displayed showing the occurrences of patents by year based on the user's selection criteria. The user has the ability to focus on a specific time period and to select various shape queries to explore the trends as described above.

The phrases that match an increasing usage query on US patents in the category "Induced Nuclear Reactions: Processes, Systems and Elements" are shown in Figure 9.6.

Evaluation

Like TOA, the presentation of PatentMiner in [LAS97] lacks an evaluation component. While it automatically generates and displays potential trends, no claim is made as to the validity of these trends. The visualization is intuitive, but no user study on its effectiveness is reported in [LAS97]. In addition, no metrics are employed in [LAS97] to verify that the trends discovered are correctly identified.

9.2.7   HDDI™

Our research has led to the development of the Hierarchical Distributed Dynamic Indexing (HDDI™) system [PKM01, BCG⁺01, BP00]. The HDDI™ system supports core text processing including information/feature extraction, feature subset selection, unsupervised and supervised text mining, and machine learning as well as evaluation for many applications, including ETD [HDD].

In [PY01] we describe our approach to the detection of emerging trends in text collections based on semantically determined clusters of terms. The HDDI™ system is used to extract linguistic features from a repository of textual data and to generate clusters based on the semantic similarity of these features. The algorithm takes a snapshot of the statistical state of a collection at multiple points in time. The rate of change in the size of the clusters and in the frequency and association of features is used as input to a neural network that classifies topics as emerging or nonemerging.

Initially we modeled the complex nonlinear classification process using neural networks. The data sets, which included three years of abstracts related to processor and pipelining patent applications, were separated by year and a set of concepts and clusters was developed for each year. In order to develop a training set, 14530 concepts were extracted and manually labeled. The system was, for example, able to correctly identify "Low power CMOS with DRAM" as an emerging trend in the proper timeframe.

In follow-on experiments we were able to duplicate the precision achieved by the neural network with the C4.5 decision tree learning algorithm [Zho00]. The run-time performance for training was significantly better with the decision tree approach. These experiments show that it is possible to detect emerging concepts in an online environment.

Like most other algorithms that we have reviewed, our approach relies on a domain expert for the final determination; thus the goal of the system is to identify emerging topics whenever possible (i.e., maximize recall) while not sacrificing precision. Unlike the first story detection algorithms, our research focuses on integrative or nondisruptive emergence of topics, as opposed to the sudden appearance of completely new topics.

Input Data and Attributes

Four databases were used to formulate a corpus in [PY01]; the US patent database, the Delphion patent database [Del], the INSPEC® database, and the COMPENDEX® database. Initial attribute selection (Table 9.10) requires parsing and tagging before extraction. The parser retains only relevant sections of the original documents. The tagger maps a part-of-speech label to each word using lexical and contextual rules [Bri92]. A finite-state machine extracts complex noun phrases

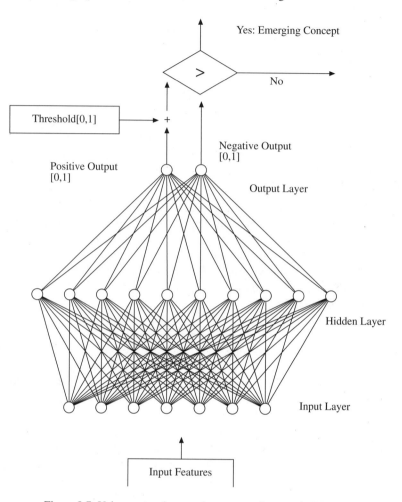

Figure 9.7. Using a neural net to detect emerging trends [PY01].

(concepts) according to the regular expression

$$C?(G|P|J) * N + (I * D?C?(G|P|J) * N+) * ,   (9.1)$$

where C is a cardinal number, G is a verb (gerund or present participle), P is a verb (past participle), J is an adjective, N is a noun, I is a preposition, D is a determiner, ? indicates zero or one occurrence, | indicates union, * indicates zero or more occurrences, and + indicates one or more occurrence [BCG+01]. Counts of each concept and counts of co-occurrence of concept pairs are recorded at this point [PY01].

An asymmetric similarity between concept pairs is calculated based on a cluster weight function described in [CL92]. The concepts are then grouped into regions of semantic locality using sLoc, an algorithm we describe in [BP00]. The maximum, mean, and standard deviation of the similarity, along with a parameter α that is a multiplication factor of the number of standard deviations, determine the threshold τ used in the first step of the sLoc algorithm. Cluster size is used in the last step (both are pruning mechanisms). As α decreases, τ increases and the number of connections between pairs of concepts decreases, resulting in smaller but more focused semantic regions. Too small a value of α could produce too many regions, while too large a value may result in only a single large region. Thus statistically finding the optimum value for α (and the threshold τ) is worthwhile, and work continues in this area. Empirical research supports an optimum value of $\alpha = 1.65$ [Yan00]. The identification of regions of semantic locality is an unsupervised learning step that produces values used as attributes in the ETD supervised learning process (see Table 9.13).

An emerging concept satisfies two principles: it should grow semantically richer over time (i.e., occur with more concepts in its region), and it should occur more often as more items reference it [PY01]. Using a cluster-based rather than an item-based approach, the artificial neural network model takes seven inputs (and one tuning threshold parameter) to classify a concept as emerging or not [PY01]. The seven inputs are described in Table 9.14.

Attribute	Detail	Generation
Regular expression	A concept (see text for definition), e.g., "emerging trend detection"	Automatic
Frequency	Number of times each concept occurs over all documents	Automatic
Frequency	Number of co-occurrences of concept pairs over all documents	Automatic
Similarity	Arc weight between concepts	Automatic
Mean	Average arc weight	Automatic
Standard deviation	Arc weight standard deviation	Automatic

Table 9.13. HDDI™ Attributes for Regions of Semantic Locality

Learning Algorithms

As mentioned above, our fundamental premise is that computer algorithms can detect emerging trends by tracing changes in concept frequency and association over time. Our approach involves separating the data into time-determined bins (as in PatentMiner and TimeMines) and taking a snapshot of the statistical relationships between terms. Two particular features were important in our model. Similar to other algorithms, the frequency of occurrence of a term should increase if the term is related to an emerging trend. Also, the term should co-occur with an increasing number of other terms if it is an emerging trend. To our knowledge, only our system has exploited term co-occurrence for automatic ETD.

Attribute	Detail	Generation
Frequency	Number of times concept occurs in trial year	Automatic
Frequency	Number of times concept occurs in the year before trial year	Automatic
Frequency	Number of times concept occurs in the year two years before trial year	Automatic
Frequency	Total number of times concept occurs in all years before trial year	Automatic
Count	Number of concepts in region containing the concept in trial year	Automatic
Count	Number of concepts in region containing the concept in the year before trial year	Automatic
Count	Number of words in the concept with length at least four	Automatic

Table 9.14. HDDI™ Attributes for Emerging Trend Detection

As noted above, the first learning model we employed is a feedforward, back-propagation artificial neural network (Figure 9.7). We used a standard three-layer network (one input layer, one hidden layer, one output layer). The number of hidden neurons was varied to optimize our results.

The attributes were extracted as described in the previous section and used as input to the neural network model [PY01], and to various other data mining algorithms such as a decision tree, support vector machine, and so on [Zho00]. In all cases, we determined that the algorithms could be trained to detect emerging trends. As with other systems, precision was fairly low (although much better than the baseline) and final determination as to whether a term displayed by the system represents an emerging trend must be left to a domain expert.

Visualization

Visualization is ongoing for trend detection within the HDDI™ system.

Evaluation

Both concept extraction and trend detection evaluations were performed. For concept extraction [BCG+01], mean precision (number of system-identified correct concepts / total number of system-identified concepts) and mean recall (number of system-identified correct concepts / total number of human expert-identified concepts) were calculated for several collections. Two of the test collections (drawn from the Grainger DLI database [UIU], the US Patent Office, and the aforementioned commercial patent database Delphion) had precision ranges of [95.1, 98.7] and [95.2, 99.2], respectively, and recall ranges of [77.4, 91.3] and [75.6, 90.6], respectively, with 95% confidence.

Automatic trend detection performance in the HDDI™ system was measured by precision, recall, and F_β [PY01]. An average precision of 0.317 constituted a 4.62 factor of improvement over random baseline precision; recall averaged 0.359. Either metric could be improved by altering the neural network threshold parameter. Since good recall was the primary focus, F_β, a weighted average of precision and recall with parameter β, was also examined. β is the precision weight and

$$F_\beta = \frac{(1 + \beta^2) \cdot precision \cdot recall}{\beta^2 \cdot precision + recall}. \tag{9.2}$$

9.2.8   Other Related Work

[FD95] proposes a technique for development of a hierarchical data structure from text databases. This data structure then facilitates the study of concept distributions in text. The authors propose comparing the concept distributions from adjacent time periods. This approach to trend analysis seems promising; however, we were not able to obtain a more detailed description of the approach, or the experimental results, so we are unable to present a more comprehensive summary. Feldman has also been active in the development of commercial products for emerging trend detection (Section 9.3).

We have focused on research efforts that identify trends based primarily on the use of words and phrases; however, several research groups are using a different approach. [CC99, PFL+00, Ley02] present algorithms that primarily employ citation information for trend detection.

Several systems focus more on the visualization of textual data and can be adapted to trend detection at the discretion of the user. One such system, Envision [NFH+96], allows users to explore trends graphically in digital library metadata (including publication dates) to identify emerging concepts. It is basically a multimedia digital library of computer science literature, with full-text searching and full-content retrieval capabilities. The system employs the use of colors and shapes to convey important characteristics of documents. For example, the interface uses color to show the degree of relevance of a document.

Plaisant et al. describe a visual environment called LifeLines for reviewing personal medical histories in [PMS+98]. The visualization environment presented in their work exploits the timeline concept to present a summary view of patient data.

The EAnalyst [LSL+00] analyzes two types of data, textual and numeric, both with timestamps. The system predicts trends in numeric data based on the content of textual data preceding the trend. For example, the system predicts the trend in stock prices based on articles published prior to the appearance of the (numeric) trend.

9.3   Commercial Software Overview

Commercial software products are available to aid a company interested in ETD. Some companies specialize in providing content [Lex, Mor, Nor, IDC, Gar, Fac], some provide general-purpose information retrieval capabilities such as feature extraction [App, Cap, Loc], document management [HyB, Ban], search and retrieval [Ban, HyB, Lex, Tex], categorization [INS, Int, Sem, Ser, SPSb, Str, Ver], and clustering [Auta, Cle, INS, Tex, Tho]. Although all of these products can be used to facilitate an ETD effort, only a few have capabilities specifically geared toward trend analysis and detection. These products are briefly introduced in this section.

9.3.1   Autonomy

The Clusterizer™ tool provided by Autonomy [Auta] provides support for ETD. Clusterizer™ [Clu] takes a corpus as input and produces a set of clusters of information from the data. These clusters can be used to identify new topics in the data by taking a set of clusters from a previous time period and comparing it to a set of clusters in the current period. In essence, Clusterizer™ displays a view of clusters over time. The tool is thus designed to show trends in clusters, including the appearance and disappearance of clusters, as well as changes in cluster size.

The pattern matching algorithms of nonlinear adaptive digital signal-processing, Claude Shannon's principles of information theory, Bayesian inference, and neural networks form the core of Autonomy's technology [Aut99]. Since natural language contains much duplication, concepts that are repeated less frequently in a document are assumed to correspond to the essence of that document. Autonomy thus describes a document with patterns based on usage and frequency of terms. Adaptive probabilistic concept modeling is used to determine relevance between documents and further train the system [Autb].

The Autonomy Clusterizer™ module assists a human domain expert in detecting trends. The Breaking News pane automatically finds new clusters of information by comparing clusters from different time periods. The Spectrograph pane is a visualization tool that plots clusters as lines over time. The color and width of the lines indicate the size and quality of the cluster; changes signal an increase or decrease in significance.

Like the research tools described in the previous section, Spectrograph is designed to provide the user with a view into the data. The domain expert must then use the data to form conclusions as to the validity of a given trend. In terms of evaluation, to the best of our knowledge no formal assessment has been conducted of the performance of these tools when used for ETD.

9.3.2   SPSS LexiQuest

LexiQuest products use advanced natural language processing technology to access, manage, and retrieve textual information [SPSb]. LexiQuest LexiMine is a text mining tool designed to help the user obtain new insights by identifying key concepts and the relationships between them. It employs a combination of dictionary-based linguistic analysis and statistical proximity matching to identify key concepts (i.e., terms) as well as the degree of relationship between concepts [Bry02].

Concept identification is achieved through the use of unabridged dictionaries and thesauri that contain (multiword) terms used to match terms in LexiMine's input text. Term occurrences and term co-occurrences are counted either by paragraph or document and are used to build relational concept maps. Although no machine learning algorithms are employed per se, LexiQuest's term similarity formulas are akin to those used in association mining [Bry02].

The concept relationships are portrayed in a graphical map that displays the cumulative occurrence of concepts. The map can be utilized to investigate trends. Further analysis can be achieved by importing LexiMine data into the related Clementine [SPSa] tool. As with many of the research and commercial tools discussed in this survey, the validity of trends is ultimately left to a human domain expert and tool performance is neither quantified nor evaluated in any formal way [Bry02].

9.3.3   ClearForest

ClearForest provides a platform and products to extract relevant information from large amounts of text and to present summary information to the user [Cle]. Two products, ClearResearch and ClearSight, are useful for ETD applications. Clear-Research is designed to present a single-screen view of complex interrelationships, enabling users to view news and research content in context. ClearSight provides simple graphic visualizations of relationships among companies, people, and events in the business world. It also provides real-time updates of new product launches, management changes, emerging technologies, and so on in any specified context. Users can drill down further into each topic to view more information or read related articles [Gra02].

ClearForest uses a rule-based approach to identify "entities" and "facts." An entity is a sequence of one or more words corresponding to a single concept. A fact is a relationship between entities. Rules can contain part-of-speech or stem identifiers for words, references to dictionaries and lexicons, and structural characteristics. The extraction engine applies rules to input documents and outputs tagged information. The precise location of all information extracted is recorded. Occurrence and co-occurrence of concepts are used by the analytic engine to summarize information. Visualization tools display this information in various levels of detail [Gra02].

ClearForest divides the detection process into stages. The first, gathering information, is performed by search engines. The second and third, extracting and consolidating information, are managed by ClearForest. The fourth, identifying a valid trend, is handled by a human domain expert using ClearForest's Trends Graph display. In terms of evaluation, like LexiQuest, ClearForest has not performed formal evaluation of the ETD component [Gra02].

9.4   Conclusions and Future Work

We have described several semiautomatic and fully automatic ETD systems, providing detailed information related to linguistic and statistical features, training and test set generation, learning algorithms, visualization, and evaluation. This review of the literature indicates that much progress has been made toward automating the process of detecting emerging trends, but there remains room for improvement. All of the systems surveyed rely on human domain expertise to separate emerging trends from noise in the system. As a result, research projects that focus on creating effective processes to both semi- and fully automatically detect emerging trends, develop effective visualizations, and apply various learning algorithms to assist with ETD can and should continue.

In addition, we discovered that few systems, whether research or commercial in nature, have employed formal evaluation metrics and methodologies to determine effectiveness. The development and use of metrics for evaluation of ETD systems is critical. The results published to date simply do not allow us to compare systems to one another. In a step designed to address this issue we are in the process of building the HDDI™ textual data mining software infrastructure that includes algorithms for formal evaluation of ETD systems (see `http://hddi.cse.lehigh.edu`).

Wider use of the TDT [TDT] data sets will also be helpful in the process of standardizing evaluation of ETD systems. In addition, usability studies need to be conducted for visualization systems. Additional training sets geared specifically toward trend detection also need to be developed. Toward this end, we have developed a back-end to our CIMEL system (Section 9.2.2) that gathers data generated by students who use the ETD component of CIMEL. This will aid us in developing techniques to automatically generate training sets for use in machine learning approaches to ETD.

We also note that projects tend to focus either on applying machine learning techniques to trend detection, or on the use of visualization techniques. Both techniques, when used alone, have proved inadequate thus far. Techniques that blend the use of visualization with machine learning may hold more promise. As a result, we are extending our HDDI™ system to include a visualization component for trend detection. Early prototypes hold promise, but, as noted above, usability studies must be conducted to prove the effectiveness of our approach.

A final point: to the best of our knowledge, no formal studies have been conducted of the (manual) processes employed by domain experts in ETD. Such a study would employ standard tools such as surveys and focus groups to develop a (manual) methodology for ETD.[3] We plan to pursue the execution of such a study in the near term in order to define a robust methodology that can be automated using extensions of the various techniques employed in our previous work [BCG+01, BPK+01, BPK+02, BP00, PY01, PCP01, PKM01, RGP02].

9.5   Industrial Counterpoint: Is ETD Useful?   Dr. Daniel J. Phelps, Leader, Information Mining Group, Eastman Kodak

The Information Mining Group at Eastman Kodak Company has been following developments in the text mining field since 1998. Initially, our interest was in using text mining tools to help us do a better job of understanding the content of patents. More recently, we have expanded our interest to include mining science and technology literature. We have had practical experience identifying suitable data sources, working with both custom and commercial tools, and presenting information in a form that our clients find useful. This background gives me a good perspective for commenting on the potential usefulness of Emerging Trend Detection (ETD) tools and some of the challenges that will arise in trying to use them in the corporate environment.

The objective of ETD is to provide an automated alert when new developments are happening in a specific area of interest. It is assumed that a detected trend is an indication that some event has occurred. The person using the ETD software will look at the data to determine the underlying development. Whether the development is important is a judgment call that depends on the situation and the particular information needs of the person evaluating the data.

The need to become aware of new developments in science, technology, or business is critical to decision makers at all levels of a corporation. These people need to make better data-driven decisions as part of their daily work. They need data that are complete and available in a timely manner. Traditionally, people have learned about a majority of the new developments by reading various types of text documents or by getting the information from others who have read the documents. As the pace of new developments accelerates and the number of documents increases exponentially, it will no longer be possible for an individual to keep up with what is happening by using manual processes. There is a clear need for new tools and methodologies to bring some level of automation to detect trends and new developments. ETD tools have the potential to play an important role in identifying new developments for corporate decision makers. These tools should help make it possible to look through more data sources for new developments and do it in less time than with current manual methods.

---

[3]Note that in [RGP02] we developed such a methodology, which has been partially automated in the CIMEL system (Section 9.2.2).

To better understand what capabilities an ETD tool must have to be useful, one has to look at who will be using the tool. There are several broad groups of potential users in a corporation: the first group is the analysts or information professionals who work to fulfill the information needs of others; the second is the individual contributors looking for information relevant to their own projects; the third is the managers who need to make strategic and/or tactical decisions.

Analysts work with information as the main component of their jobs. These people work on projects specified by clients. The output of a given project will be a report delivered to the client for use in the decision-making process. Analysts are trained in information retrieval techniques, text mining techniques, and so on, and are familiar with the various information sources needed to complete a given project. They have to be able to communicate the results of the work in a form clients can easily use and understand. Taking time to learn new tools and methodologies is an expected part of the job.

An ETD tool that is targeted for use by analysts can be complex. The analysts will have to be given sufficient training to become proficient in its use. Because the analysts will use the tool for multiple projects, they will learn the capabilities and limitations of the tool and be able to recognize those areas where its application is appropriate. One would expect a sophisticated user interface that would allow analysts to access the relevant data sources, process the underlying text, and display the results in a meaningful way using computer graphics visualization techniques. The visualization scheme used must draw the analysts' attention to trends and allow them to drill down into the data to find out what developments lead to what trends. The determination of whether a detected trend is important is complicated by the fact that the analysts do not always know what clients will judge to be important. Interaction between the analysts and the clients is critical to ensure that clients' needs are met. This is typically an iterative process as the analysts learn more about what information the clients need, and the clients find out what information is actually available. Once the analysis is done, the ETD tool should have the ability to export information to facilitate report generation.

The scientists, engineers, or business people who want to use ETD tools to obtain project-specific information need a tool that is easy to learn and intuitive to use. Connecting to the appropriate data sources and processing the data must be transparent to the users. This user group will typically have limited training in the tool and will use it only occasionally. They will not have the time to learn all the nuances of using the software. The information that is used will be that which is delivered automatically. A simple, graphical user interface with easily interpreted graphical visualizations is required. These people have the advantage that they are performing the work for themselves, therefore, they can make the determination whether newly detected trends are actually important.

An ETD tool meant to be used by management personnel must automatically be connected to the appropriate data sources, have an intuitive user interface, be very easy to learn, and provide output in a consistent format with which the managers are comfortable. Extremely sophisticated visualizations that are difficult to interpret and require high levels of interaction will not be useful in this environment. In the current corporate culture, managers do not have the time to engage in anything but the most cursory training in new tools. This means they will probably not be able to operate the ETD tool effectively enough to complete the analysis themselves. They are generally more comfortable with having an analyst assemble the information and provide an executive summary. The summarized information and the underlying data could be presented to them using a browser format that would allow them to look at the high-level results and then go further into detail when they find something in which they are particularly interested.

No matter how capable an ETD tool becomes, the quality of the output will depend upon the quality of the input data. Because ETD tools are supposed to identify new developments, the data processed through the tool must be current. There are several news services, such as Factiva [Fac], that supply all types of news on a continuous basis. Currently, there are no equivalent services for fundamental science and technology information. One has to search a variety of sources to find the required information. The sampling frequency used to extract the data from the newsfeed needs to reflect the rapidity with which things change in the area. Business and financial events happen much more frequently than changes in technology. One might set up the ETD tool to collect news data each day and process them to look for new trends. Since developments in science and technology occur at a slower pace, it might be appropriate to work in blocks of one week or one month. Processing information in one-year blocks is adequate for a retrospective look at what has happened, but it is not acceptable for most decision-making situations.

ETD systems will have to be customizable to meet the needs of specific clients. The ETD algorithms will eventually become proficient at determining when something has happened. However, whether the development is important depends on the needs of the person who is looking at the data. Broadly speaking, there are two types of client profile information that must be obtained: the definition of the area of interest, and the definition of what is characterized as an important development in that area. Traditional alerting systems handle the first problem by setting up a user profile containing a search query that is run on a periodic basis against specified data sources. Typically, the query is built on a trial-and-error basis by the information professional and the client. This is an iterative process. Some of the newer knowledge-management systems use training sets to determine the characteristics of the documents of interest and build the "query." Each new document is checked against the target characteristics and a decision is automatically made as to whether the document belongs to the area of interest. What remains to be seen is which approach will work best with a given ETD tool. Either process can take a significant amount of time for the client who wants the information. There is also the problem that the client's areas of interest will expand and change over time. Each time this happens, an existing profile will have to be modified or a new profile will have to be generated.

The problem of determining what is a significant event in a given area is handled in interactive systems by having the decision makers operate the tools themselves. If the decision makers are unable or unwilling to work directly with the tool, analysts will have to interview the decision makers and obtain the basic guidelines with which to work. The analysts will perform the analysis and compile a list of potentially significant developments for the decision maker's review. It would be best if this processed information were presented in a browser format that would allow the decision makers to drill down into the detail underlying any development they find to be of interest.

It is too early to predict the cost of a commercial ETD software system. If it is comparable to the knowledge management and text database mining software of today, it will cost tens of thousands to hundreds of thousands of dollars. It probably will carry the same sort of fee structure as the high-end software packages available today. Vendors charge an initial license purchase price and require an annual maintenance fee to provide technical support and updates of the software. Sometimes it is possible to buy the software individually by the "seat," but often the vendors push to sell a corporate license. If only a few people will be using the software, then purchasing seats makes sense. If the software is actually going to be used across the enterprise, then a corporate license is probably the better choice. Another cost that is often overlooked is the impact on the corporate IT infrastructure. There can be a capital cost to purchase the high-performance hardware needed to run calculation-intensive ETD applications. Even when the application is run on existing inhouse servers, there is usually the need to have a system administrator, and possibly a database administrator, available to keep the application up and running.

To get a picture of what an ETD tool might look like in the future, it is helpful to examine a perfect-world scenario for an executive information system that would include an ETD capability. The characteristics of such a scenario are depicted in Table 9.15.

1. Raw data processed into useful information
2. Sufficient information presented to meet current need
3. No irrelevant information presented
4. All information available immediately when needed
5. Information prioritized for the current need
6. Information presented in a format that is intuitively easy to understand
7. Information can be viewed at different levels of detail
8. Information can be viewed from multiple perspectives

Table 9.15. Perfect-World Scenario

Executive decision makers are extremely busy and want to make good data-driven decisions as fast as possible. This means they cannot take the time to assemble and process the raw data themselves. They want complete, timely, processed information, sorted in an appropriate prioritized order for the decisions at hand. They do not want to waste time looking at redundant or irrelevant information. The information needs to be presented in a format that is intuitively easy to understand and can be looked at in different levels of detail and from multiple perspectives. An excellent ETD tool will have these same characteristics and will meet the needs of all three groups of potential users in the corporation.

There continues to be good progress made in knowledge management and text mining tools. Because ETD systems make use of these types of tools, I think there is a good possibility that practical ETD systems will eventually become available for fixed-information needs. Building a system that will keep up with a given decision maker's changing information needs will be difficult, unless a good method is found to automatically translate the subject areas of interest and the important developments criteria from the words of the user to the ETD system. It will always be a challenge to assure that data sources available for processing are adequate to support the needs of the decision maker.

In this section, I have reviewed some of the practical aspects of working with an ETD tool in a corporate environment. The real test for an ETD system is whether it provides useful information about new developments to the decision maker in a timely manner. The current systems do not seem to provide the required level of performance to be used extensively in the corporate environment. There is hope that new generations of ETD tools will be useful to corporate decision makers when they become available.

Acknowledgments: It is with genuine gratitude that we thank the following individuals: Michael Berry, editor of this volume and Chair of the 2001 SIAM Text Mining Workshop, and Doug Bryan at SPSS (LexiQuest) and Barry Graubart at ClearForest for providing timely information included in our commercial products section. We also wish to acknowledge the National Science Foundation for their role in funding the CIMEL project (Section 9.2.2) and our HDDI™ textual data mining software infrastructure [HDD] under Grant Number 0087977. Finally, coauthor William M. Pottenger would like to express his sincere gratitude to his Lord and Savior, Jesus Christ, for His continuing help in his life.

References

[ABC+95]   J. Allan, L. Ballesteros, J. Callan, W. Croft, and Z. Lu. Recent experiments with inquery. In *Proceedings of the Fourth Text Retrieval Conference (TREC-4)*, pages 49–63, 1995.

[APL98]   J. Allan, R. Papka, and V. Lavrenko. On-line new event detection and tracking. In *Proceedings of ACM SIGIR*, pages 37–45, 1998.

[App]   Applied Semantics [online, cited July 2002]. Available from World Wide Web: www.appliedsemantics.com.

[APWZ95]   R. Agrawal, G. Psaila, E.L. Wimmers, and M. Zait. Querying shapes of histories. In *Proceedings of the 21st International Conference on Very Large Databases*, Zurich, Sep 1995.

[AS95]   R. Agrawal and R. Srikant.Mining sequential patterns.In *Proceedings of the International Conference on Data Engineering (ICDE)*, Taipei, Mar 1995.

[Auta]   Autonomy [online, cited July 2002].Available from World Wide Web: www. autonomy.com.

[Autb]   Autonomy [online, cited July 2002].Available from World Wide Web: www. autonomy.com/Content/Technology/Background/ IntellectualFoundations.

[Aut99]   Knowlege Suite (Review) [online].1999 [cited July 2002].Available from World Wide Web: www.autonomy.com/Extranet/Marketing/ Analyst White Papers/Butler Report on Autonomy Suite 200299.pdf.

[Ban]   Banter [online, cited July 2002].Available from World Wide Web: www. banter.com.

[BCG⁺01]   R. Bader, M. Callahan, D. Grim, J. Krause, N. Miller, and W.M. Pottenger.The role of the HDDI™ collection builder in hierarchical distributed dynamic indexing.In *Proceedings of the Textmine '01 Workshop, First SIAM International Conference on Data Mining*, Apr 2001.

[BMSW97]   D. Bikel, S. Miller, R. Schwartz, and R. Weischedel.Nymble: A high-performance learning name-finder.In *Proceedings of the Fifth Conference on Applied Natural Language Processing*, pages 194–201, 1997.

[BP00]   F. Bouskila and W.M. Pottenger.The role of semantic locality in hierarchical distributed dynamic indexing.In *Proceedings of the 2000 International Conference on Artificial Intelligence (IC-AI 2000)*, Las Vegas, Jun 2000.

[BPK⁺01]   G.D. Blank, W.M. Pottenger, G.D. Kessler, M. Herr, H. Jaffe, S. Roy, D. Gevry, and Q. Wang.Cimel: Constructive, collaborative inquiry-based multimedia e-learning.In *Proceedings of the Sixth Annual Conference on Innovation and Technology in Computer Science Education (ITiCSE)*, Jun 2001.

[BPK⁺02]   G.D. Blank, W.M. Pottenger, G.D. Kessler, S. Roy, D.R. Gevry, J.J. Heigl, S.A. Sahasrabudhe, and Q. Wang.Design and evaluation of multimedia to teach Java and object-oriented software engineering.*American Society for Engineering Education*, Jun 2002.

[Bri92]   E. Brill.A simple rule-based part of speech tagger.In *Proceedings of the Third Conference on Applied Natural Language Processing*. ACL, 1992.

[Bry02]   D. Bryan, Jul 2002.Email correspondence.

[Cap]   Captiva [online, cited July 2002].Available from World Wide Web: www. captivacorp.com.

[CC99]   C. Chen and L. Carr.A semantic-centric approach to information visualization.In *Proceedings of the 1999 International Conference on Information Visualization*, pages 18–23, 1999.

[CIM]   CIMEL [online, cited July 2002].Available from World Wide Web: www. cse.lehigh.edu/~cimel.

[CL92]   H. Chen and K.J. Lynch.Automatic construction of networks of concepts characterizing document databases.*IEEE Transactions on Systems, Man and Cybernetics*, 22(5):885–902, 1992.

[Cle]   ClearForest [online, cited July 2002].Available from World Wide Web: `www.`
            `clearforest.com`.

[Clu]   Clusterizer™ [online, cited July 2002].Available from World Wide Web:
            `www.autonomy.com/Extranet/Technical/Modules/`
            `TB Autonomy Clusterizer.pdf`.

[COM]   COMPENDEX® [online, cited July 2002].Available from World Wide Web:
            `edina.ac.uk/compendex`.

[Del]   Delphion [online, cited July 2002].Available from World Wide Web: `www.`
            `delphion.com`.

[DHJ⁺98]   G.S. Davidson, B. Hendrickson, D.K. Johnson, C.E. Meyers, and B.N.
            Wylie.Knowledge mining with VxInsight™: Discovery through interac-
            tion.*Journal of Intelligent Information Systems*, 11(3):259–285, 1998.

[Edg95]   E. Edgington.*Randomization Tests*.Marcel Dekker, New York, 1995.

[Fac]   Factiva [online, cited July 2002].Available from World Wide Web: `www.`
            `factiva.com`.

[FD95]   R. Feldman and I. Dagan.Knowledge discovery in textual databases.In *Pro-
            ceedings of the First International Conference on Knowledge Discovery
            (KDD-95)*. ACM, New York, Aug 1995.

[FSM⁺95]   D. Fisher, S. Soderland, J. McCarthy, F. Feng, and W. Lehnert.Description of
            the UMASS systems as used for MUC-6.In *Proceedings of the Sixth Message
            Understanding Conference*, pages 127–140, Nov 1995.

[Gar]   GartnerG2 [online, cited July 2002].Available from World Wide Web: `www.`
            `gartnerg2.com/site/default.asp`.

[Gev02]   D. Gevry.Detection of emerging trends: Automation of domain expert prac-
            tices.Master's thesis, Department of Computer Science and Engineering at
            Lehigh University, 2002.

[Gra02]   B. Graubart.White paper, turning unstructured data overload into a competitive
            advantage, Jul 2002.Email attachment.

[HDD]   HDDI™ [online, cited July 2002].Available from World Wide Web: `hddi.`
            `cse.lehigh.edu`.

[HHWN02]   S. Havre, E. Hetzler, P. Whitney, and L. Nowell.ThemeRiver: Visualiz-
            ing thematic changes in large document collections.*IEEE Transactions on
            Visualization and Computer Graphics*, 8(1), Jan – Mar 2002.

[HyB]   HyBrix [online, cited July 2002].Available from World Wide Web: `www.`
            `siemens.com/index.jsp`.

[IDC]   IDC [online, cited July 2002].Available from World Wide Web: `www.idc.`
            `com`.

[INS]   INSPEC® [online, cited July 2002].Available from World Wide Web: `www.`
            `iee.org.uk/Publish/INSPEC`.

[Int]   Interwoven [online, cited July 2002].Available from World Wide Web: `www.`
            `interwoven.com/products`.

[LA00a]   A. Leuski and J. Allan.Lighthouse: Showing the way to relevant information.In
            Proceedings of the IEEE Symposium on Information Visualization (InfoVis),
            pages 125–130, 2000.

[LA00b]   A. Leuski and J. Allan.Strategy-based interactive cluster visualization for information retrieval.*International Journal on Digital Libraries*, 3(2):170–184, 2000.

[LAS97]   B. Lent, R. Agrawal, and R. Srikant.Discovering trends in text databases.In *Proceedings of the Third International Conference on Knowledge Discovery and Data Mining*, Newport Beach, CA, pages 227–230, 1997.

[Lex]   LexisNexis [online, cited July 2002].Available from World Wide Web: www.lexisnexis.com.

[Ley02]   L. Leydesdorff.Indicators of structural change in the dynamics of science: Entropy statistics of the sci journal citation reports.*Scientometrics*, 53(1):131–159, 2002.

[Lin]   Linguistic Data Consortium [online, cited July 2002].Available from World Wide Web: www.ldc.upenn.edu.

[Loc]   Lockheed-Martin [online, cited July 2002].Available from World Wide Web: www.lockheedmartin.com.

[LSL⁺00]   V. Lavrenko, M. Schmill, D. Lawrie, P. Ogilvie, D. Jensen, and J. Allan.Mining of concurrent text and time-series.In *Proceedings of the ACM KDD-2000 Text Mining Workshop*, 2000.

[MDOP97]   A. Martin, T.K.G. Doddington, M. Ordowski, and M. Przybocki.The DET curve in assessment of detection task performance.In *Proceedings of EuroSpeech '97*, vol. 4, pages 1895–1898, 1997.

[Mor]   Moreover [online, cited July 2002].Available from World Wide Web: www.moreover.com.

[NFH⁺96]   L.T. Nowell, R.K. France, D. Hix, L. S Heath, and E.A. Fox.Visualizing search results: Some alternatives to query-document similarity.In *Proceedings of SIGIR'96*, Zurich, pages 67–75, 1996.

[Nor]   Northern Light [online, cited July 2002].Available from World Wide Web: www.northernlight.com.

[PCP01]   W.M. Pottenger, M.R. Callahan, and M.A. Padgett.Distributed information management.*Annual Review of Information Science and Technology (ARIST)*, 35, 2001.

[PD95]   A.L. Porter and M.J. Detampel.Technology opportunities analysis.*Technological Forecasting and Social Change*, 49:237–255, 1995.

[PFL⁺00]   A. Popescul, G.W. Flake, S. Lawrence, L. Ungar, and C.L. Giles.Clustering and identifying temporal trends in document databases.In *Proceedings of IEEE Advances in Digital Libraries*, pages 173–182, 2000.

[PKM01]   W.M. Pottenger, Y. Kim, and D.D. Meling.HDDI™: Hierarchical distributed dynamic indexing.In *Data Mining for Scientific and Engineering Applications*, Robert Grossman, Chandrika Kamath, Vipin Kumar and Raju Namburu, eds., Jul 2001.

[PMS⁺98]   C. Plaisant, R. Mushlin, A. Snyder, J. Li, D. Heller, and B. Shneiderman.Lifelines: Using visualization to enhance navigation and analysis of patient records.In *Proceedings of the 1998 American Medical Informatic Association Annual Fall Symposium*, pages 76–80, 1998.

[PY01]   W.M. Pottenger and T. Yang.Detecting emerging concepts in textual data min-
            ing.In *Computational Information Retrieval*, M.W. Berry, ed., pages 89–105,
            SIAM, Philadelphia, 2001.

[RGP02]   S. Roy, D. Gevry, and W.M. Pottenger.Methodologies for trend detection in
            textual data mining.In *Proceedings of the Textmine '02 Workshop, Second
            SIAM International Conference on Data Mining*, Apr 2002.

[Roy02]   S. Roy.A multimedia interface for emerging trend detection in inquiry-based
            learning.Master's thesis, Department of Computer Science and Engineering
            at Lehigh University, May 2002.

[SA96]   R. Srikant and R. Agrawal.Mining sequential patterns: Generalizations
            and performance improvements.In *Proceedings of the Fifth International
            Conference on Extending Database Technology (EDBT)*, Avignon, 1996.

[SA00]   R. Swan and J. Allan.Automatic generation of overview timelines.In *Proceed-
            ings of the 23rd Annual International ACM SIGIR Conference on Research
            and Development in Information Retrieval*, Athens, ACM, New York, pages
            49–56, 2000.

[Sem]   Semio [online, cited July 2002].Available from World Wide Web: www.
            semio.com.

[Ser]   Ser Solutions [online, cited July 2002].Available from World Wide Web: www.
            sersolutions.com.

[SJ00]   R. Swan and D. Jensen.TimeMines: Constructing timelines with statistical
            models of word usage.In *Proceedings of the Sixth ACM SIGKDD International
            Conference on Knowledge Discovery and Data Mining*, 2000.

[SPSa]   SPSS Clementine [online, cited July 2002].Available from World Wide Web:
            www.spss.com/spssbi/clementine.

[SPSb]   SPSS LexiQuest [online, cited July 2002].Available from World Wide Web:
            www.spss.com/spssbi/lexiquest.

[Str]   Stratify [online, cited July 2002].Available from World Wide Web: www.
            stratify.com.

[TDT]   TDT [online, cited July 2002].Available from World Wide Web: www.nist.
            gov/speech/tests/tdt/index.htm.

[Tex]   TextAnalyst [online, cited July 2002].Available from World Wide Web: www.
            megaputer.com/products/ta/index.php3.

[Tho]   ThoughtShare [online, cited July 2002].Available from World Wide Web:
            www.thoughtshare.com.

[UIU]   University of Illinois at Urbana-Champaign Digital Library Initiative [online,
            cited July 2002].Available from World Wide Web: dli.grainger.uiuc.
            edu.

[US]   US Patent Site [online, cited July 2002].Available from World Wide Web:
            www.uspto.gov/main/patents.htm.

[Ver]   Verity [online, cited July 2002].Available from World Wide Web: www.
            verity.com.

[XBC94]   J. Xu, J. Broglio, and W.B. Croft.The design and implementation of a part
            of speech tagger for English.Technical report, Center for Intelligent Informa-

          tion Retrieval, University of Massachusetts, Amherst, Technical Report IR-52, 1994.

[Yan00]   T. Yang.Detecting emerging conceptual contexts in textual collections.Master's thesis, Department of Computer Science at the University of Illinois at Urbana-Champaign, 2000.

[YPC98]   Y. Yang, T. Pierce, and J. Carbonell.A study on retrospective and on-line event detection.In *Proceedings of SIGIR-98, 21st ACM International Conference on Research and Development in Information Retrieval*, 1998.

[Zho00]   L. Zhou.Machine learning classification for detecting trends in textual collections.Master's thesis, Department of Computer Science at the University of Illinois at Urbana-Champaign, December 2000.

Bibliography

[ABC⁺95]   J. Allan, L. Ballesteros, J. Callan, W. Croft, and Z. Lu.Recent experiments with inquery.In *Proceedings of the Fourth Text Retrieval Conference (TREC-4)*, pages 49–63, 1995.

[AD91]   H. Almuallim and T.G. Dietterich.Learning with many irrelevant features.In *Proceedings of the Ninth National Conference on Artificial Intelligence*, pages 547–552, 1991.

[AL01]   R. Ando and L. Lee.Latent semantic space.In *Proceedings of the ACM Special Interest Group for Information Retrieval (SIGIR) Conference*, Helsinki, Finland, pages 154–162, 2001.

[And00]   R. Ando.Latent semantic space.In *Proceedings of the ACM Special Interest Group for Information Retrieval (SIGIR) Conference*, Athens, pages 216–223, 2000.

[APL98]   J. Allan, R. Papka, and V. Lavrenko.On-line new event detection and tracking.In *Proceedings of ACM SIGIR*, pages 37–45, 1998.

[App]   Applied Semantics [online, cited July 2002].Available from World Wide Web: www.appliedsemantics.com.

[APWZ95]   R. Agrawal, G. Psaila, E.L. Wimmers, and M. Zait.Querying shapes of histories.In *Proceedings of the 21st International Conference on Very Large Databases*, Zurich, Sep 1995.

[AS95]   R. Agrawal and R. Srikant.Mining sequential patterns.In *Proceedings of the International Conference on Data Engineering (ICDE)*, Taipei, Mar 1995.

[Auta]   Autonomy [online, cited July 2002].Available from World Wide Web: www.autonomy.com.

[Autb]   Autonomy [online, cited July 2002].Available from World Wide Web: www.autonomy.com/Content/Technology/Background/IntellectualFoundations.

[Aut99]   Knowlege Suite (Review) [online].1999 [cited July 2002].Available from World Wide Web: www.autonomy.com/Extranet/Marketing/Analyst White Papers/Butler Report on Autonomy Suite 200299.pdf.

[Ban]   Banter [online, cited July 2002].Available from World Wide Web: www.banter.com.

[BB99]   M.W. Berry and M. Browne.*Understanding Search Engines: Mathematical Modeling and Text Retrieval*.SIAM, Philadelphia, 1999.

[BB02]   P. Berkhin and J.D. Becher.Learning simple relations: Theory and applica-
            tions.In *Proceedings of the Second SIAM International Conference on Data
            Mining*, Arlington, VA, pages 410-436, April 2002.

[BCG+01]   R. Bader, M. Callahan, D. Grim, J. Krause, N. Miller, and W.M. Pot-
            tenger.The role of the HDDI™ collection builder in hierarchical distributed
            dynamic indexing.In *Proceedings of the Textmine '01 Workshop, First SIAM
            International Conference on Data Mining*, Apr 2001.

[BDJ99]   M. Berry, Z. Drmač, and E. Jessup.Matrices, vector spaces, and information
            retrieval.*SIAM Review*, 41(2):335–362, 1999.

[BDO95]   M. Berry, S. Dumais, and G. O'Brien.Using linear algebra for intelligent
            information retrieval.*SIAM Review*, 37(4):573–595, 1995.

[BE99]   R. Barzilayand and M. Elhadad.*Using Lexical Chains for Text Summariza-
            tion.*In [MM99], 1999.

[Bez81]   J.C. Bezdek.*Pattern Recognition with Fuzzy Objective Function Algo-
            rithms.*Plenum, New York, 1981.

[BF98]   P.S. Bradley and U.M. Fayyad.Refining initial points for K-Means clus-
            tering.In *Procedings of the Fifteenth International Conference on Machine
            Learning*, Morgan Kaufmann, San Francisco, pages 91–99, 1998.

[BFR98]   P.S. Bradley, U.M. Fayyad, and C. Reina.Scaling clustering algorithms to
            large databases.In *Knowledge Discovery and Data Mining*, pages 9–15, 1998.

[BGG+99a]   D. Boley, M. Gini, R. Gross, E.-H. Han, K. Hastings, G. Karypis, V. Kumar,
            B. Mobasher, and J. Moore.Document categorization and query generation
            on the World Wide Web using WebACE.*AI Review*, 13(5,6):365–391, 1999.

[BGG+99b]   D. Boley, M. Gini, R. Gross, E.-H. Han, K. Hastings, G. Karypis, V. Kumar,
            B. Mobasher, and J. Moore.Partitioning-based clustering for Web document
            categorization.*Decision Support Systems*, 27(3):329–341, 1999.

[Bjö96]   Å. Björck.*Numerical Methods for Least Squares Problems.*SIAM, Philadel-
            phia, 1996.

[BL85]   C. Buckley and A.F. Lewit.Optimizations of inverted vector searches.In *SIGIR
            '85*, pages 97–110, 1985.

[BMSW97]   D. Bikel, S. Miller, R. Schwartz, and R. Weischedel.Nymble: A high-
            performance learning name-finder.In *Proceedings of the Fifth Conference
            on Applied Natural Language Processing*, pages 194–201, 1997.

[Bol98]   D.L. Boley.Principal direction divisive partitioning.*Data Mining and Knowl-
            edge Discovery*, 2(4):325–344, 1998.

[Bow]   Bow: A toolkit for statistical language modeling, text retrieval, classification
            and clustering [online, cited September 2002].Available from World Wide
            Web: www.cs.cmu.edu/~mccallum/bow.

[BP98]   S. Brin and L. Page.The anatomy of a large-scale hypertextual Web search
            engine.*Computer Networks and ISDN Systems*, 30(1–7):107–117, 1998.

[BP00]   F. Bouskila and W.M. Pottenger.The role of semantic locality in hierarchi-
            cal distributed dynamic indexing.In *Proceedings of the 2000 International
            Conference on Artificial Intelligence (IC-AI 2000)*, Las Vegas, Jun 2000.

[BPK+01]   G.D. Blank, W.M. Pottenger, G.D. Kessler, M. Herr, H. Jaffe, S. Roy,
            D. Gevry, and Q. Wang.Cimel: Constructive, collaborative inquiry-based

multimedia e-learning.In *Proceedings of the Sixth Annual Conference on Innovation and Technology in Computer Science Education (ITiCSE)*, Jun 2001.

[BPK+02]   G.D. Blank, W.M. Pottenger, G.D. Kessler, S. Roy, D.R. Gevry, J.J. Heigl, S.A. Sahasrabudhe, and Q. Wang.Design and evaluation of multimedia to teach Java and object-oriented software engineering.*American Society for Engineering Education*, Jun 2002.

[BR01]   K. Blom and A. Ruhe.Information retrieval using very short Krylov sequences.In *Proceedings of the Computational Information Retrieval Conference held at North Carolina State University, Raleigh, Oct. 22, 2000*, M. Berry, ed., SIAM, Philadelphia, pages 39–52, 2001.

[Bri92]   E. Brill.A simple rule-based part of speech tagger.In *Proceedings of the Third Conference on Applied Natural Language Processing*. ACL, 1992.

[Bry02]   D. Bryan, Jul 2002.Email correspondence.

[BS01]   V.D. Blondel and P.P. Senellart.Automatic extraction of synonyms in a dictionary.Technical Report 89, Université catholique de Louvain, Louvain-la-neuve, Belgium, 2001.Presented at the Text Mining Workshop 2002 in Arlington, VA.

[BV02]   V.D. Blondel and P. Van Dooren.A measure of graph similarity between graph vertices.Technical Report, Université catholique de Louvain, Louvain-la-neuve, Belgium, 2002.

[BYRN99]   R. Baeza-Yates and B. Ribeiro-Neto.*Modern Information Retrieval*.Addison-Wesley, Boston, 1999.

[Cap]   Captiva [online, cited July 2002].Available from World Wide Web: `www.captivacorp.com`.

[CC99]   C. Chen and L. Carr.A semantic-centric approach to information visualization.In *Proceedings of the 1999 International Conference on Information Visualization*, pages 18–23, 1999.

[CF79]   R.E. Cline and R.E. Funderlic.The rank of a difference of matrices and associated generalized inverses.*Linear Algebra Appl.*, 24:185–215, 1979.

[CFG95]   M.T. Chu, R.E. Funderlic, and G.H. Golub.A rank-one reduction formula and its applications to matrix factorizations.*SIAM Review*, 37(4):512–530, 1995.

[CHJ61]   W.D. Climenson, H.H. Hardwick, and S.N. Jacobson.Automatic syntax analysis in machine indexing and abstracting.*American Documentation*, 12(3):178–183, 1961.

[CIM]   CIMEL [online, cited July 2002].Available from World Wide Web: `www.cse.lehigh.edu/~cimel`.

[CKPT92]   D.R. Cutting, D.R. Karger, J.O. Pedersen, and J.W. Turkey.Scatter/gather: A cluster-based approach to browsing large document collections.In *Proceedings of the Fifteenth Annual International ACM SIGIR Conference on Research and Development in Information Retrieval*, Copenhagen, Denmark, pages 318–329, Jun 1992.

[CL92]   H. Chen and K.J. Lynch.Automatic construction of networks of concepts characterizing document databases.*IEEE Transactions on Systems, Man and Cybernetics*, 22(5):885–902, 1992.

[Cle]   ClearForest [online, cited July 2002].Available from World Wide Web: `www.clearforest.com`.

[Clu]   Clusterizer™ [online, cited July 2002].Available from World Wide Web: `www.autonomy.com/Extranet/Technical/Modules/TB Autonomy Clusterizer.pdf`.

[CMU]   20 newsgroup data set [online, cited September 2002].Available from World Wide Web: `www-2.cs.cmu.edu/afs/cs.cmu.edu/project/theo-20/www/data/news20.html`.

[COM]   COMPENDEX® [online, cited July 2002].Available from World Wide Web: `edina.ac.uk/compendex`.

[Cro90]   C.J. Crouch.An approach to the automatic construction of global thesauri.*Information Processing and Management*, 26:629–640, 1990.

[Dam95]   M. Damashek.Gauging similarity with n-grams: Language-independent categorization of text.*Science*, 267:843–848, 1995.

[DDF$^+$90] S. Deerwester, S. Dumais, G. Furnas, T. Landauer, and R. Harshman.Indexing by latent semantic analysis.*Journal of the American Society for Information Science*, 41(6):391–407, 1990.

[De02]   I. Dhillon and J. Kogan (eds.).*Proceedings of the Workshop on Clustering High Dimensional Data and its Applications*.SIAM, Philadelphia, 2002.

[Del]   Delphion [online, cited July 2002].Available from World Wide Web: `www.delphion.com`.

[Dem97]   J. Demmel.*Applied Numerical Linear Algebra*.SIAM, Philadelphia, 1997.

[DGK02]   I.S. Dhillon, Y. Guan, and J. Kogan.Refining clusters in high-dimensional text data.In *Proceedings of the Workshop on Clustering High Dimensional Data and Its Applications at the Second SIAM International Conference on Data Mining*, I.S. Dhillon and J. Kogan, eds., pages 71–82. SIAM, Philadelphia, 2002.

[DH99]   J. Dean and M.R. Henzinger.Finding related pages in the World Wide Web.*WWW8 / Computer Networks*, 31(11-16):1467–1479, 1999.

[DHJ$^+$98] G.S. Davidson, B. Hendrickson, D.K. Johnson, C.E. Meyers, and B.N. Wylie.Knowledge mining with VxInsight™: Discovery through interaction.*Journal of Intelligent Information Systems*, 11(3):259–285, 1998.

[DHS01]   R.O. Duda, P.E. Hart, and D.G. Stork.*Pattern Classification*, second edition.Wiley, New York, 2001.

[Die97]   T.G. Dietterich.Machine-learning research: Four current directions.*AI Magazine*, 18(4):97–136, 1997.

[DM01]   I.S. Dhillon and D.S. Modha.Concept decompositions for large sparse text data using clustering.*Machine Learning*, 42(1):143–175, Jan 2001.Also appears as IBM Research Report RJ 10147, Jul 1999.

[DMK02]   I.S. Dhillon, S. Malella, and R. Kumar.Enhanced word clustering for hierarchical text classification.In *KDD-2002*, 2002.

[DPHS98]   S. Dumais, J. Platt, D. Heckerman, and M. Sahami.Inductive learning algorithms and representations for text categorization.In *Proceedings of the ACM CIKM International Conference on Information and Knowledge Management*, Bethesda, MD, Nov 1998.

[Edg95]   E. Edgington.*Randomization Tests.*Marcel Dekker, New York, 1995.

[Edm68]   H.P. Edmundson.New methods in automatic extraction.*Journal of the ACM*, 16(2):264–285, 1968.

[EE98]   M.A. Elmi and M. Evens.Spelling correction using context.In *Proceedings of the 36th Annual Meeting of the ACL and the 17th International Conference on Computational Linguistics*, pages 360–364, 1998.

[EY39]   C. Eckart and G. Young.A principal axis transformation for non-Hermitian matrices.*Bulletin of the American Mathematics Society*, 45:118–121, 1939.

[Fac]   Factiva [online, cited July 2002].Available from World Wide Web: www.factiva.com.

[FD95]   R. Feldman and I. Dagan.Knowledge discovery in textual databases.In *Proceedings of the First International Conference on Knowledge Discovery (KDD-95)*. ACM, New York, Aug 1995.

[FK97]   H. Frigui and R. Krishnapuram.Clustering by competitive agglomeration.*Pattern Recognition*, 30(7):1223–1232, 1997.

[FK99]   H. Frigui and R. Krishnapuram.A robust competitive clustering algorithm with applications in computer vision.*IEEE Transactions on Pattern Analysis and Machine Intelligence*, 21(5):450–465, May 1999.

[FLE00]   F. Farnstrom, J. Lewis, and C. Elkan.Scalability for clustering algorithms revisited.*SIGKDD Explorations*, 2(1):51–57, 2000.

[FN00]   H. Frigui and O. Nasraoui.Simultaneous clustering and attribute discrimination.In *Proceedings of the IEEE Conference on Fuzzy Systems*, San Antonio, TX, pages 158–163, 2000.

[For65]   E. Forgy.Cluster analysis of multivariate data: Efficiency vs. interpretability of classifications.*Biometrics*, 21(3):768, 1965.

[FS94]   E.A. Fox and J.A. Shaw.Combination of multiple sources.In *Proceedings of the Second Text Retrieval Conference (TREC-2)*, pages 97–136, 1994.

[FSM$^+$95] D. Fisher, S. Soderland, J. McCarthy, F. Feng, and W. Lehnert.Description of the UMASS systems as used for MUC-6.In *Proceedings of the Sixth Message Understanding Conference*, pages 127–140, Nov 1995.

[Fuk90]   K. Fukunaga.*Introduction to Statistical Pattern Recognition*, second edition.Academic, Boston, MA, 1990.

[Gar]   GartnerG2 [online, cited July 2002].Available from World Wide Web: www.gartnerg2.com/site/default.asp.

[GBNP96]   R.L. Grossman, H. Bodek, D. Northcutt, and H.V. Poor.Data mining and tree-based optimization.In *Proceedings of the Second International Conference on Knowledge Discovery and Data Mining*, E. Simoudis, J. Han and U. Fayyad, eds., AAAI Press, Menlo Park, CA, pages 323–326, 1996.

[Gev02]   D. Gevry.Detection of emerging trends: Automation of domain expert practices.Master's thesis, Department of Computer Science and Engineering at Lehigh University, 2002.

[GK79]   E.E. Gustafson and W.C. Kessel.Fuzzy clustering with a fuzzy covariance matrix.In *Proceedings of the IEEE Conference on Decision and Control*, San Diego, pages 761–766, 1979.

[GK02]   E. Gendler and J. Kogan.Index terms selection for clustering large text data.In *Proceedings of the Workshop on Text Mining at the Second SIAM International Conference on Data Mining*, M.W. Berry, ed., pages 87–94, 2002.

[GKSR⁺02] M. Ganapathiraju, J. Klein-Seetharaman, R. Rosenfeld, J. Carbonell, and R. Reddy.Rare and frequent n-grams in whole-genome protein sequences.In *Proceedings of RECOMB'02: The Sixth Annual International Conference on Research in Computational Molecular Biology*, 2002.

[GV96]   G. Golub and C. Van Loan.*Matrix Computations*, third edition.John Hopkins Univ. Press, Baltimore, MD, 1996.

[GL02]   R.L. Grossman and R.G. Larson.A state space realization theorem for data mining. In subm., 2002.

[Gra02]   B. Graubart.White paper, turning unstructured data overload into a competitive advantage, Jul 2002.Email attachment.

[Gre93]   G. Grefenstette.Automatic thesaurus generation from raw text using knowledge-poor techniques.In *Making Sense of Words. Ninth Annual Conference of the UW Centre for the New OED and Text Research. 9*, 1993.

[Gre94]   G. Grefenstette.*Explorations in Automatic Thesaurus Discovery*.Kluwer Academic, Boston, 1994.

[Gre01]   E. Greengrass.Information retrieval: A survey.United States Department of Defense Technical Report TR-R52-008-001, 2001.

[Gut57]   L. Guttman.A necessary and sufficient formula for matric factoring.*Psychometrika*, 22(1):79–81, 1957.

[Har95]   D.K. Harman, editor.*Proceedings of the Third Text Retrieval Conference (TREC-3)*. National Institute of Standards and Technology Special Publication 500-226, 1995.

[Har99]   D. Harman.Ranking algorithms.In *Information Retrieval*, R. Baeza-Yates and B. Ribeiro-Neto, eds., ACM, New York, pages 363–392, 1999.

[Hay99]   S. Haykin.*Neural Networks: A comprehensive foundation, second edition*. Prentice-Hall, Upper Saddle River, NJ, 1999.

[HDD]   HDDI™ [online, cited July 2002].Available from World Wide Web: hddi. cse.lehigh.edu.

[Hey01]   M. Heymans.Extraction d'information dans les graphes, et application aux moteurs de recherche sur internet, Jun 2001.Université Catholique de Louvain, Faculté des Sciences Appliquées, Département d'Ingénierie Mathématique.

[HHWN02]   S. Havre, E. Hetzler, P. Whitney, and L. Nowell.ThemeRiver: Visualizing thematic changes in large document collections.*IEEE Transactions on Visualization and Computer Graphics*, 8(1), Jan – Mar 2002.

[HJP03]   P. Howland, M. Jeon, and H. Park.Structure preserving dimension reduction for clustered text data based on the generalized singular value decomposition.*SIAM Journal on Matrix Analysis and Applications*, 2003, to appear.

[HMH00]   L. Hubert, J. Meulman, and W. Heiser.Two purposes for matrix factorization: A historical appraisal.*SIAM Review*, 42(1):68–82, 2000.

[HOB99]   L.O. Hall, I.O. Ozyurt, and J.C. Bezdek.Clustering with a genetically optimized approach.*IEEE Transactions on Evolutionary Computations*, 3(2):103–112, Jul 1999.

[Hor65]   P. Horst.*Factor Analysis of Data Matrices*.Holt, Rinehart and Winston, Orlando, FL, 1965.

[Hot33]   H. Hotelling.Analysis of a complex of statistical variables into principal components.*Journal of Educational Psychology*, 24:417–441, 1933.

[HL99]   E.H. Hovy and H. Liu.The value of indicator phrases for automated text summarization.Unpublished, 1999.

[HP02]   P. Howland and H. Park.Extension of discriminant analysis based on the generalized singular value decomposition.Technical Report 021, Department of Computer Science and Engineering, University of Minnesota, Twin Cities, 2002.

[HPS96]   D.A. Hull, J.O. Pedersen, and H. Schütze.Method combination for document filtering.In *Proceedings of the Nineteenth Annual International ACM SIGIR Conference on Research and Development in Information Retrieval*, 1996.

[Hub81]   P.J. Huber.*Robust Statistics*.Wiley, New York, 1981.

[HyB]   HyBrix [online, cited July 2002].Available from World Wide Web: www. siemens.com/index.jsp.

[IDC]   IDC [online, cited July 2002].Available from World Wide Web: www.idc. com.

[INS]   INSPEC® [online, cited July 2002].Available from World Wide Web: www. iee.org.uk/Publish/INSPEC.

[Int]   Interwoven [online, cited July 2002].Available from World Wide Web: www. interwoven.com/products.

[Inx]   Inx [online].Available from World Wide Web: www.inxight.com/ products/linguistx.

[JD88]   A. Jain and R. Dubes.*Algorithms for Clustering Data*.Prentice-Hall, Englewood Cliffs, NJ, 1988.

[JKP94]   G. John, R. Kohavi, and K. Pfleger.Irrelevant features and the subset selection problem.In *Proceedings of the Eleventh International Machine Learning Conference*, pages 121–129, 1994.

[JP01]   B.J. Jansen and U. Pooch.A review of Web searching studies and a framework for future research.*Journal of the American Society for Information Science and Technology*, 52(3):235–246, 2001.

[JSS00]   B.J. Jansen, A. Spink, and T. Saracevic.Real life, real users, and real needs: A study and analysis of user queries on the Web.*Information Processing and Management*, 36(2):207–227, 2000.

[JW99]   J. Jannink and G. Wiederhold.Thesaurus entry extraction from an on-line dictionary.In *Proceedings of Fusion '99*, Sunnyvale, CA, Jul 1999.

[KA02]   M. Kobayashi and M. Aono.Major and outlier cluster analysis using dynamic re-scaling of document vectors.In *Proceedings of the SIAM Text Mining Workshop, Arlington, VA*, SIAM, Philadelphia, pages 103–113, 2002.

[KAST01]   M. Kobayashi, M. Aono, H. Samukawa, and H. Takeuchi.*Information retrieval apparatus for accurately detecting multiple outlier clusters*.patent, filing, IBM Corporation, 2001.

[KAST02]   M. Kobayashi, M. Aono, H. Samukawa, and H. Takeuchi.Matrix computations for information retrieval and major and outlier cluster detection.*Journal of Computational and Applied Mathematics*, 149(1):119–129, 2002.

[Kat96]   S. Katz.Distribution of context words and phrases in text and language modeling.*Natural Language Engineering*, 2(1):15–59, 1996.

[KHKL96]   T. Kohonen, J. Hynninen, J. Kangas, and J. Laaksonen.Som_pak: The self-organizing map program package.*Laboratory of Computer and Information Science*, Report A31, 1996.

[KK93]   R. Krishnapuram and J. M. Keller.A possibilistic approach to clustering.*IEEE Transactions on Fuzzy Systems*, 1(2):98–110, May 1993.

[Kle99]   J.M. Kleinberg.Authoritative sources in a hyperlinked environment.*Journal of the ACM*, 46(5):604–632, 1999.

[KMS00]   M. Kobayashi, L. Malassis, and H. Samukawa.*Retrieval and ranking of documents from a database*.patent, filing, IBM Corporation, 2000.

[Kog01a]   J. Kogan.Clustering large unstructured document sets.In *Computational Information Retrieval*, M.W. Berry, ed., pages 107–117, SIAM, Philadelphia, 2001.

[Kog01b]   J. Kogan.Means clustering for text data.In *Proceedings of the Workshop on Text Mining at the First SIAM International Conference on Data Mining*, M.W. Berry, ed., pages 47–57, 2001.

[Kog02]   J. Kogan.Computational information retrieval.*Springer-Verlag Lecture Notes in Contributions to Statistics*, H.R. Lerche, ed., 2002. To appear.

[Koh92]   T. Kohonen.*The Self-Organizing Map. Neural Networks: Theoretical Foundations and Analysis*.IEEE Press, New York, 1992.

[Kor77]   R.R. Korfhage.*Information Storage and Retrieval*.Wiley, New York, 1977.

[Kow97]   G. Kowalski.*Information Retrieval Systems: Theory and Implementation*. Kluwer Academic, Hingham, MA, 1997.

[KPC95]   J. Kupied, J. Piedersen, and F. Chen.A trainable document summarizer.In *Proceedings of the Eighteenth Annual International SIGIR Conference on Research and Development in Information Retrieval*, pages 68–73, 1995.

[KR92]   K. Kira and L. A. Rendell.The feature selection problem: Traditional methods and a new algorithm.In *Proceedings of the Tenth National Conference on Artificial Intelligence*, pages 129–134, 1992.

[KS95]   R. Kohavi and D. Sommerfield.Feature subset selection using the wrapper model: Overfitting and dynamic search space topology.In *Proceedings of the First International Conference on Knowledge Discovery and Data Mining*, pages 192–197, 1995.

[Kue87]   G.H. Kuenning.International ispell version 3.1.00.`ftp.cs.ucla.edu`, 1987.

[Kuk92]   K. Kukich.Techniques for automatically correcting words in text. *ACM Computing Surveys*, 24(4):377–439, 1992.

[LA00a]   A. Leuski and J. Allan.Lighthouse: Showing the way to relevant informa-
tion.In *Proceedings of the IEEE Symposium on Information Visualization
(InfoVis)*, pages 125–130, 2000.

[LA00b]   A. Leuski and J. Allan.Strategy-based interactive cluster visualization for
information retrieval.*International Journal on Digital Libraries*, 3(2):170–
184, 2000.

[LAS97]   B. Lent, R. Agrawal, and R. Srikant.Discovering trends in text databases.In
*Proceedings of the Third International Conference on Knowledge Discovery
and Data Mining*, Newport Beach, CA, pages 227–230, 1997.

[Lee95]   J.H. Lee.Combining multiple evidence from different properties of weighting
schemes.In *Proceedings of the Eighteenth Annual International ACM SIGIR
Conference on Research and Development in Information Retrieval*, 1995.

[Lee97]   J.H. Lee.Analyses of multiple evidence combination.In *Proceedings of the
Twentieth Annual International ACM SIGIR Conference on Research and
Development in Information Retrieval*, 1997.

[Leh82]   W.G. Lehnert.*Plot Units: A Narrative Summarization Strategy*.Erlbaum,
Hillsdale, NJ, 1982.

[Lex]   LexisNexis [online, cited July 2002].Available from World Wide Web: www.
lexisnexis.com.

[Ley02]   L. Leydesdorff.Indicators of structural change in the dynamics of science:
Entropy statistics of the sci journal citation reports.*Scientometrics*, 53(1):131–
159, 2002.

[LH95]   C.L. Lawson and R.J. Hanson.*Solving Least Squares Problems*.SIAM,
Philadelphia, 1995.

[Lin]   Linguistic Data Consortium [online, cited July 2002].Available from World
Wide Web: www.ldc.upenn.edu.

[Loc]   Lockheed-Martin [online, cited July 2002].Available from World Wide Web:
www.lockheedmartin.com.

[LSCP96]   D. Lewis, R. Schapire, J. Cllan, and R. Papka.Training algorithms for linear
text classifiers.In *Proceedings of SIGIR-96, Nineteenth ACM International
Conference on Research and Development in Information Retrieval*, 1996.

[LSL⁺00]   V. Lavrenko, M. Schmill, D. Lawrie, P. Ogilvie, D. Jensen, and J. Al-
lan.Mining of concurrent text and time-series.In *Proceedings of the ACM
KDD-2000 Text Mining Workshop*, 2000.

[Luh58]   H.P. Luhn.The automatic creation of literature abstracts.*IBM Journal of
Research and Development*, 2(2), 1958.

[Mar97]   D. Marcu.*The Rhetorical Parsing, Summarization and Generation of Natural
Language Texts*.PhD dissertation. University of Toronto, 1997.

[May00]   J. Mayfield.Personal communication, 2000.

[McI82]   M.D. McIlroy.Development of a spelling list.*IEEE Transactions on Communication, COM-30*, 1:91–99, Jan 1982.

[MDOP97]   A. Martin, T.K.G. Doddington, M. Ordowski, and M. Przybocki.The
DET curve in assessment of detection task performance.In *Proceedings of
EuroSpeech '97*, vol. 4, pages 1895–1898, 1997.

[MKB79]   K. Mardia, J. Kent, and J. Bibby.*Multivariate Analysis.*Academic, New York, 1979.

[Mla99]   D. Mladenic.Text learning and related intelligent agents.*IEEE Expert*, Jul 1999.

[MM99]   I. Mani and M. Maybury.*Introduction. Advances in Automatic Text Summarization.*MIT Press, Cambridge, MA, 1999.

[MMP00]   J. Mayfield, P. McNamee, and C. Piatko.The JHU/APL HAIRCUT System at TREC-8.National Institute of Standards and Technology Special Publication, 2000.

[Mor]   Moreover [online, cited July 2002].Available from World Wide Web: `www.moreover.com`.

[MS00]   C. Manning and H. Schütze.*Foundations of Statistical Natural Language Processing.*MIT Press, Cambridge, MA, 2000.

[NFH$^+$96]   L.T. Nowell, R.K. France, D. Hix, L. S Heath, and E.A. Fox.Visualizing search results: Some alternatives to query-document similarity.In *Proceedings of SIGIR'96*, Zurich, pages 67–75, 1996.

[NK96]   O. Nasraoui and R. Krishnapuram.An improved possibilistic c-means algorithm with finite rejection and robust scale estimation.In *Proceedings of the North American Fuzzy Information Processing Society Conference*, Berkeley, CA, pages 395–399, Jun 1996.

[NK97]   O. Nasraoui and R. Krishnapuram.A genetic algorithm for robust clustering based on a fuzzy least median of squares criterion.In *Proceedings of the North American Fuzzy Information Processing Society Conference*, Syracuse, NY, pages 217–221, Sept 1997.

[NK00]   O. Nasraoui and R. Krishnapuram.A novel approach to unsupervised robust clustering using genetic niching.In *Proceedings of the IEEE International Conference on Fuzzy Systems*, New Orleans, pages 170–175, 2000.

[Nor]   Northern Light [online, cited July 2002].Available from World Wide Web: `www.northernlight.com`.

[Nun90]   G. Nunberg.The linguistics of punctuation.*Center for the Study of Language and Information Lecture Notes* 90(18), 1990.

[OPT00]   The online plain text english dictionary, 2000.`http://msowww.anu.edu.au/~ralph/OPTED/`.

[Ort87]   J. Ortega.*Matrix Theory: A Second Course.*Plenum, New York, 1987.

[Par97]   B. Parlett.*The Symmetric Eigenvalue Problem.*SIAM, Philadelphia, 1997.

[PAT99]   PATTERN.The pattern system version 2.6, Magnify, Inc., 1999.

[PCP01]   W.M. Pottenger, M.R. Callahan, and M.A. Padgett.Distributed information management.*Annual Review of Information Science and Technology (ARIST)*, 35, 2001.

[PCS00]   A.L. Prodromidis, P.K. Chan, and S.J. Stolfo.Meta-learning in distributed data mining systems, issues and approaches.In *Advances in Distributed Data Mining*, Hillol Kargupta and Philip Chan, eds., MIT Press, Cambridge, MA, pages 81–113, 2000.

[PD95]   A.L. Porter and M.J. Detampel.Technology opportunities analysis.*Technological Forecasting and Social Change*, 49:237–255, 1995.

[Pea01]   K. Pearson.On lines and planes of closest fit to systems of points in space.*The London, Edinburgh and Dublin Philosophical Magazine and Journal of Science, Sixth Series*, 2:559–572, 1901.

[PFL+00]   A. Popescul, G.W. Flake, S. Lawrence, L. Ungar, and C.L. Giles.Clustering and identifying temporal trends in document databases.In *Proceedings of IEEE Advances in Digital Libraries*, pages 173–182, 2000.

[PJR01]   H. Park, M. Jeon, and J. Rosen.Lower dimensional representation of text data in vector space based information retrieval.In *Proceedings of the Computational Information Retrieval Conference held at North Carolina State University, Raleigh, Oct. 22, 2000*, M. Berry, ed., SIAM, Philadelphia, pages 3–24, 2001.

[PJR03]   H. Park, M. Jeon, and J.B. Rosen.Lower dimensional representation of text data based on centroids and least squares.*BIT*, 2003, to appear.

[PKM01]   W.M. Pottenger, Y. Kim, and D.D. Meling.HDDI™: Hierarchical distributed dynamic indexing.In *Data Mining for Scientific and Engineering Applications*, Robert Grossman, Chandrika Kamath, Vipin Kumar and Raju Namburu, eds., Jul 2001.

[PMS+98]   C. Plaisant, R. Mushlin, A. Snyder, J. Li, D. Heller, and B. Shneiderman.Lifelines: Using visualization to enhance navigation and analysis of patient records.In *Proceedings of the 1998 American Medical Informatic Association Annual Fall Symposium*, pages 76–80, 1998.

[PN96]   C. Pearce and C. Nicholas.TELLTALE: Experiments in a dynamic hypertext environment for degraded and multilingual data.*Journal of the American Society for Information Science*, 47:263–275, 1996.

[Por80]   M.F. Porter.An algorithm for suffix stripping.*Program*, 14:130–137, 1980.

[PS81]   C.C. Paige and M.A. Saunders.Towards a generalized singular value decomposition.*SIAM Journal on Numerical Analysis*, 18(3):398–405, 1981.

[PY01]   W.M. Pottenger and T. Yang.Detecting emerging concepts in textual data mining.In *Computational Information Retrieval*, M.W. Berry, ed., pages 89–105, SIAM, Philadelphia, 2001.

[QOSG02]   Y. Qu, G. Ostrouchov, N. Samatova, and A. Geist.Principal component analysis for dimension reduction in massive distributed data sets.In *SIAM Workshop on High Performance Data Mining*, S. Parthasarathy, H. Kargupta, V. Kumar, D. Skillicorn, and M. Zaki, eds., Arlington, VA, pages 7–18, 2002.

[Ras92]   E. Rasmussen.Clustering algorithms.In *Information Retrieval*, W. Frakes and R. Baeza-Yates, eds., Prentice-Hall, Englewood Cliffs, NJ, pages 419–442, 1992.

[RGP02]   S. Roy, D. Gevry, and W.M. Pottenger.Methodologies for trend detection in textual data mining.In *Proceedings of the Textmine '02 Workshop, Second SIAM International Conference on Data Mining*, Apr 2002.

[RK92]   L.A. Rendell and K. Kira.A practical approach to feature selection.In *Proceedings of the International Conference on Machine Learning*, pages 249–256, 1992.

[RL87]   P.J. Rousseeuw and A.M. Leroy.*Robust Regression and Outlier Detection*.Wiley, New York, 1987.

[Roy02]   S. Roy.A multimedia interface for emerging trend detection in inquiry-based learning.Master's thesis, Department of Computer Science and Engineering at Lehigh University, May 2002.

[RR94]   J. Reynar and A. Ratnaparkhi.A maximum entropy approach to identifying sentence boundaries.In *Proceedings of the Conference on Applied Natural Language*, 1994.

[RW00]   N. Ross and D. Wolfram.End user searching on the Internet: An analysis of term pair topics submitted to the Excite Search Engine.*Journal of the American Society for Information Science and Technology*, 51(10):949–958, 2000.

[SA96]   R. Srikant and R. Agrawal.Mining sequential patterns: Generalizations and performance improvements.In *Proceedings of the Fifth International Conference on Extending Database Technology (EDBT)*, Avignon, 1996.

[SA00]   R. Swan and J. Allan.Automatic generation of overview timelines.In *Proceedings of the 23rd Annual International ACM SIGIR Conference on Research and Development in Information Retrieval*, Athens, ACM, New York, pages 49–56, 2000.

[Sal71]   G. Salton.*The SMART Retrieval System*.Prentice-Hall, Englewood Cliffs, NJ, 1971.

[Sal89]   G. Salton.*Automatic Text Processing: The Transformation, Analysis, and Retrieval of Information by Computer*.Addison-Wesley, Reading, MA, 1989.

[SBC97]   B. Shneiderman, D. Byrd, and W.B. Croft.Clarifying search: A user-interface framework for text searches.*D-Lib Magazine*, 1:1–18, 1997.

[Sem]   Semio [online, cited July 2002].Available from World Wide Web: www. semio.com.

[Sen01]   P. P. Senellart.Extraction of information in large graphs. Automatic search for synonyms.Technical Report 90, Université catholique de Louvain, Louvain-la-neuve, Belgium, 2001.

[Ser]   Ser Solutions [online, cited July 2002].Available from World Wide Web: www.sersolutions.com.

[SHMM99]   C. Silverstein, M. Henzinger, H. Marais, and M. Moricz.Analysis of a very large Web search engine query log.*SIGIR Forum*, 33(1):6–12, 1999.

[SJ00]   R. Swan and D. Jensen.TimeMines: Constructing timelines with statistical models of word usage.In *Proceedings of the Sixth ACM SIGKDD International Conference on Knowledge Discovery and Data Mining*, 2000.

[Ska94]   D. Skalak.Prototype and feature selection by sampling and random mutation hill climbing algorithms.In *Proceedings of the Eleventh International Machine Learning Conference (ICML-94)*, pages 293–301, 1994.

[SKK00]   M. Steinbach, G. Karypis, and V. Kumar.A comparison of document clustering algorithms.In *Proceedings of the KDD Workshop on Text Mining*, 2000.

[SM83]   G. Salton and M.J. McGill.*Introduction to Modern Information Retrieval*.Mc Graw-Hill, New York, 1983.